Zeta-functions:
An introduction to
algebraic geometry

A D Thomas

University College of Swansea

Zeta-functions: An introduction to algebraic geometry

Pitman

LONDON · SAN FRANCISCO · MELBOURNE

PITMAN PUBLISHING LIMITED
39 Parker Street, London WC2B 5PB

FEARON PITMAN PUBLISHERS INC.
6 Davis Drive, Belmont, California 94002, USA

Associated Companies
Copp Clark Pitman, Toronto
Pitman Publishing New Zealand Ltd, Wellington
Pitman Publishing Pty Ltd, Melbourne

First published 1977
Reprinted 1978

AMS Subject Classifications: (main) 14G10, 14A10, 32C10, 55C20
                                           (subsidiary) 14F20, 30A46, 42A25, 58A10

Library of Congress Cataloging in Publication Data

Thomas, Alan David, 1946 –
  Zeta-functions.

  (Research notes in mathematics; 12)
  Bibliography: p.
  Includes index.
  1. Geometry, Algebraic. 2. Functions, Zeta.
I. Title. II. Series.
QA564.T45   516′.35   76-57938
ISBN 0-273-01038-7

Reproduced and printed by photolithography
in Great Britain at Biddles of Guildford

Preface

In November 1973, Professor J.P. Serre gave two seminars at King's College, London, entitled 'Algebraic varieties over finite fields'. Stimulated by such a clear and concise exposition, I gave a post-graduate seminar course at the University College of Swansea in the following year, from which this book has developed.

I should like to thank my colleagues at Swansea for their interest and help in the preparation of the manuscript, Mrs. S.P. Campbell, Mrs. E. Jenkins and Mrs. G.M. Maddocks for their typing, and my wife Anthea for her constant encouragement.

University College of Swansea   Alan Thomas

January 1977

Contents

Introduction

The aim of this book is to explain the development of the statements and proofs of the Weil conjectures from the analogous results for the Riemann zeta function, at the same time describing the transformation of classical algebraic number theory and algebraic geometry into the language of schemes. In order to motivate the cohomological aspects and in particular the idea of an étale map, a certain amount of classical algebraic geometry (over the field of complex numbers) and associated complex manifold theory is included, not only to provide an insight to the more abstract methods, but also to be used in the various comparison techniques.

All these methods and results are contained in existing literature, which to the non-expert may seem both inpenetrable and disjointed. This text is designed to be read by any mathematician who is conversant with the basics of commutative algebra, as in Atiyah & Macdonald (17), field and Galois theory, as in Winter (217), and who has some familiarity with manifolds and algebraic topology (see for example Matsushima (138) and Spanier (183)) and categories as in Mitchell (141).

It would be impossible to cover the contents of this book with full details in a small number of pages, and so footnotes have been included to give explicit references to the relevant information. I have tried to choose the facts which have been included to give the reader the flavour of the subject. Following these footnotes, the reader will soon realise the volume of mathematics which has been omitted: I should like to think that the value of this book lies as much in what is omitted as in what is included. The decision on the precise contents was a most difficult one. Probably the most important omissions are the details and precise statement of the Riemann-Roch theorem and its implications in the context of these notes. The reason for not including this is that, in order to convey the full weight of this theorem, one would have to discuss Jacobians in more detail, and the associated K-theories, characteristic classes and index theorems which would have been impossible in the short number of pages. This possibly will be the contents of a companion volume.

We start with Riemann's zeta-function and its properties, especially those which express number theoretic properties of the ring of integers (which after all was the reason Riemann introduced it). This immediately generalises to the zeta-function of an algebraic number field, which contains similar information about the field, or more precisely, about its ring of (algebraic) integers.

Next we discuss the contents of Artin's thesis in which he generalises these classical ideas, replacing by the euclidean domain $F_p[X]$ of polynomials over the finite field of order p, and considering quadratic extensions of the associated field of

ii

fractions. This analogous situation has much in common with the
classical case and there is a natural definition of a zeta-function. Artin observed in the many cases he calculated that the
analogue of the Riemann hypothesis was true.

In order to present these two parallel theories in a unified
way, we introduce (real) valuations, global fields and the
associated functional analysis of adèles and idèles.

In Chapter 6 we show how a global field of non-zero characteristic can be interpreted geometrically as a curve over a
finite field. This then provides a link between number theory
and geometry. Replacing the finite field by the field of complex numbers we have the classical algebraic theory, and from
this the analytic theory, of Riemann surfaces, which we discuss
in Chapter 7. The analytic theory of course depends on using the
'usual' or 'strong' topology on the field of complex numbers
which is not defined algebraically. We find ourselves in the
position of being able to prove certain algebraic/geometric
results analytically but not algebraically, which is unsatis-
factory for two reasons. Firstly for aesthetic reasons,
algebraic results should have algebraic proofs, and secondly,
such analytic methods do not generalise to the study of curves
over other fields. In this context, the Appendix 7.8 is of great
significance, as it contains the essential ideas behind the
effectiveness and significance of the etale topology, and the
importance of elliptic curves or more generally abelian and
Jacobian varieties, which we discuss in Chapter 8. This Chapter

ends with Hasse's proof of the Riemann hypothesis for curves of
genus 1.

Historically, this places us at around 1936, when the curve
is still identified with its field of functions and is of no
geometrical significance. In order to prove the Riemann hypothesis for global fields of non-zero characteristic, Weil
introduced (geometrically) products of curves which lead on to
the abstract theory of algebraic varieties over arbitrary fields.

Chapter 9 starts with the definition of an affine variety,
and the concept of smoothness is shown to be an appropriate generalisation of integral closure. In order to piece together
affine varieties to obtain the general variety, we introduce the
language of sheaves, first used for this purpose by Serre (166).
This is unfortunately rather technical, but simultaneously
enables us to define the concept of a manifold, complex manifold
and variety (and, in Chapter 12, of a scheme). In §9.11 we
describe briefly the properties of cycles on a variety, whose
importance becomes apparent in Chapter 10. At this point we are
in a position to state the Weil conjectures, all of which have
now been proved using etale topology and cohomology (the fundamental properties of which are contained in the 1583 pages of
SGA 4, see (14)).

Besides stating these conjectures, Weil suggested that they
should admit a cohomological proof and outlined the lines such a
proof might take. Such a cohomology appeared around 1959/60 and
gradually the various necessary properties were established.

iv

Working analytically, Dwork proved the first of these conjectures (rationality) in 1960, and in 1962 Lubkin proved cohomologi-
cally all the conjectures (except the Riemann hypothesis) under
certain restrictions, although his proof referred to other
theorems which had themselves not been proved.

In order to motivate these cohomological attacks on the
problems, we describe in Chapter 10 how one can prove the analogous conjectures (except the Riemann hypothesis) for smooth
complex varieties, by giving them the usual topology and then
applying Čech cohomology methods. The fundamental result is the
Lefschetz fixed point theorem, which connects the number of
fixed points of a continuous map with its effect in cohomology,
and which is used along with properties of cycles and their
cohomological invariants.

Chapter 11 continues this analogy by showing that the statement and proof of the Riemann hypothesis is quite a different
problem, needing the force of the Hodge theory of harmonic
integrals to finally overcome it.

Hopefully by now the reader has reached Chapter 12, in
which the language of schemes, Grothendieck topologies and etale
maps are introduced, motivated by the earlier chapters, and we
finish with a brief account of the proofs of the Weil conjectures.

Looking back, the reader should then realise that the
Riemann hypothesis and Weil conjectures, apart from being of
intrinsic interest and direct application, have generated through
the attempts at their verification a considerable amount of other

mathematics which we would perhaps otherwise not have discovered.

To some extent then, this book is like a jigsaw puzzle. At
first selected pieces fit together. As one progresses one finds
individual pieces which do not seem to belong anywhere, but the
further one goes, the more important these pieces become, until
finally they all fit together to form the overall picture. Of
course one might say this is not the way to present a mathematical topic; a mathematical treatise should be linearly ordered.
While this is possible even desirable for 'elementary' branches
(i.e. those using a small vocabulary), it would seem to be
impossible for the contents of these notes, if one wishes to
provide the motivation and intuition. The ordering here is
loosely based on the chronology of the subject. Certain questions may be partially answered in one section, only to be
resurrected in a later section and tackled again with the aid of
techniques developed in the meantime. One final comment:
because of this approach, this book should not be read once!

NOTATION

The notation used in this book is either standard or explained
in the text, in which case the reader is referred to the index.
The symbols N, Z, Q, R, C and H denote (respectively) the
natural, integral, rational, real, complex and quaternionic
numbers, and S^n denotes the n-sphere.

All rings are assumed to be commutative with an identity
unless otherwise stated.   For any ring R we write R[X] and R(X)

for the rings of polynomial and rational functions of X over R,
and $R[[X]]$ and $R((X))$ for the rings of power and (finite)
Laurent series. If S is a subset of R, $<S>$ denotes the ideal
it generates.

For any field K we write char(K) for its characteristic,
K^a for its algebraic closure and K, $\mu(K)$, $\mu_n(K)$ for the
groups of units, roots of unity, n-th roots of unity in K, and
Gal(L/K) for the Galois group of the normal extension L/K.
F_q denotes the finite field of order q.

Indexing sets, especially for Σ, Π, $\varprojlim$ and $\varinjlim$ are
often ommited when they are obvious or irrelevant from the
context.

1 Dirichlet series

1.1   DIRICHLET SERIES

Let Ω denote the set of functions from N to C.   If we define addition of such functions componentwise, and multiplication by the formula

$$\alpha . \beta(n) \;=\; \sum_{ij=n} \alpha(i)\,\beta(j) \quad \text{for} \quad \alpha, \beta \in \Omega$$

then Ω is a commutative ring with 1, which is called the <u>ring of Dirichlet series</u> (over C).   (It is also called the ring of arithmetic functions; it is a unique factorisation domain, being isomorphic with a ring of power series[1].)   If α is a Dirichlet series, we define its <u>derivative</u> $\alpha' \in \Omega$ by the formula

$$\alpha'(n) \;=\; -\alpha(n)\log(n).$$

The function $d: \Omega \to \Omega, \;\; \alpha \to \alpha'$ is called <u>differentiation</u>.

   If z is a complex number, let $\alpha_s(z)$ denote the series $\sum_n \alpha(n)/n^z$ and if this series is convergent, let $\hat{\alpha}(z)$ denote its sum.

   By abuse of notation, we shall identify α with α_s.

1.2   CONVERGENCE OF DIRICHLET SERIES

For any real number r, let H_r denote the open half-plane $\{z \in C: \text{Re}(z) > r\}$.   Suppose that α is a Dirichlet series and z_0 is a complex number such that $\alpha_s(z_0)$ converges.   Let $r_0 = \text{Re}(z_0)$.   The series $\alpha_s(z)$ converges[2] for any $z \in H_{r_0}$, and converges uniformly[3] in any compact subspace of H_{r_0}.   Since the function $1/n^z$ is holomorphic in the complex plane, it follows that $\hat{\alpha}(z)$ is holomorphic in H_{r_0}.   If, in addition, there is a complex number z_1 such that $\alpha_s(z_1)$ converges absolutely, then[4] $\alpha_s(z)$

converges absolutely for any $z \in H_{r_1}$ where $r_1 = \mathrm{Re}(z_1)$.

Let $\Omega_a = \{\alpha \in \Omega : \exists \, z \in C \text{ such that } \alpha_s(z) \text{ converges absolutely}\}$. This is a subring of Ω, called the <u>ring of absolutely convergent Dirichlet series</u>. Let Ω_0 be the ring of (germs of)[5] holomorphic functions defined in some open half-plane, and let $j : \Omega_a \to \Omega_0$ be the function $j(\alpha) = \hat{\alpha}$. This function is a homomorphism, and it commutes with differentiation. In fact it is a monomorphism, and the proof[6] depends on the following lemma[7]:

<u>Lemma 1.2.1</u> *Let* $\mathrm{x,c}$ *be positive real numbers.   Then*

$$\frac{1}{2\pi i} \int_{c-i\infty}^{c+i\infty} \frac{x^z}{z} \, dz = \begin{cases} 1 & if \ \ x < 1 \\ 0 & if \ \ x > 1. \end{cases}$$

Suppose then that $\hat{\alpha}$ is absolutely convergent in H_r, and let $\mathrm{x,c}$ be positive real numbers such that $c > r$ and x is non-integral. We see that

$$\int_{c-i\infty}^{c+i\infty} \frac{\hat{\alpha}(z)x^z}{z} \, dz = \int_{c-i\infty}^{c+i\infty} \sum_n \frac{\alpha(n)x^z}{z\,n^z} \, dz = \sum_n \alpha(n) \int_{c-i\infty}^{c+i\infty} \frac{(x/n)^z}{z} \, dz \ ,$$

the interchange of integration and summation being possible since the series $\alpha_s(z)$ is uniformly convergent.   Thus applying 1.2.1 we obtain the equation

$$\frac{1}{2\pi i} \int_{c-i\infty}^{c+i\infty} \frac{\hat{\alpha}(z)x^z}{z} \, dz = \sum_n^{[x]} \alpha(n),$$

which is called <u>Perron's formula.</u>   This formula shows that the holomorphic function $\hat{\alpha}$ determines the arithmetic function α, so that $j : \Omega_a \to \Omega_0$ is indeed injective.   If f is a function holomorphic in some half plane and $f = j(\alpha)$ for some (unique) α, then we say that α is the <u>Dirichlet series</u> <u>of</u> f.

2

1.3 EXAMPLES OF DIRICHLET SERIES

The canonical example of a Dirichlet series is the Riemann zeta-function. This has Dirichlet series $\Sigma 1/n^z$, and is denoted by $\zeta(z)$. This Dirichlet series converges absolutely in H_1, but does not converge at $z = 1$. It is well-known[8] that the ζ-function extends to a meromorphic function on the complex plane with a simple pole (residue 1) at $z = 1$. It satisfies the functional equation:

$$(1.3.1) \qquad \zeta(z) \;=\; 2(2\pi)^{z-1}\Gamma(1-z)\sin(\pi z/2)\zeta(1-z),$$

which by means of the equations

$$\Gamma(z)\Gamma(1-z) \;=\; \pi\cosec(\pi z) \quad \text{and}$$

$$\Gamma(z)\Gamma(z+1/2) \;=\; 2^{1-2z}(\pi)^{1/2}\Gamma(2z)$$

may be rewritten

$$\pi^{-z/2}\Gamma(z/2)\zeta(z) \;=\; \pi^{-(1-z)/2}\Gamma((1-z)/2)\zeta(1-z).$$

The following proposition is generic in these notes[9].

__Proposition 1.3.2__ *For $z \in H_1$, the infinite product*

$$(1.3.2.1) \qquad \prod_{\substack{p \,\in\, N \\ p \text{ prime}}} (1 - p^{-z})^{-1}$$

converges to $\zeta(z)$. Moreover, this infinite product converges absolutely in H_1, and uniformly on compact subsets.

The fact that the value of this infinite product is $\zeta(z)$ is essentially that (formally)

$$\prod_{\substack{p \,\in\, N \\ p \text{ prime}}} (1 - p^{-z})^{-1} \;=\; \prod_{\substack{p \,\in\, N \\ p \text{ prime}}} \sum_{k=1}^{\infty} p^{-kz} \;=\; \sum_{n=1}^{\infty} n^{-z}$$

since every positive integer is uniquely expressible as a product of positive prime integers. To convert this into an analytic argument, one has to work with finite sets of primes and take the limit. This proposition is thus an an analytic statement of unique factorisation in the ring of integers.

<u>Corollary 1.3.3</u>.　*The set of prime integers is infinite.*

<u>Proof</u>.　If there were only a finite set of primes, then the product (1.3.2.1) would be a finite product of functions, each holomorphic in H_0, and so would define a holomorphic continuation of the ζ-function to H_0, which is impossible because of the pole at $z = 1$.

Since $(1 - p^{-z})^{-1}$ does not vanish in H_1, it follows from Hurwitz's theorem[10] and 1.3.2 that the ζ-function has no zeros in H_1. From the functional equation (1.3.1), it follows that the only zeros of the ζ-function in the region $\{z: \mathrm{Re}(z) < 0\}$ are the zeros of $\sin(\pi z/2)$ which occur at the points $z = -2k$ for $k \in N$. These zeros are called the <u>trivial zeros</u>. The remaining zeros thus lie in the 'critical strip' $\{z \in C: 0 \le \mathrm{Re}(z) \le 1\}$, and are called the <u>non-trivial zeros</u>.

<u>The Riemann Hypothesis (1.3.4)</u>.　*The zeros in the critical strip all lie on the line* $\mathrm{Re}(z) = 1/2$.

Since first stated by Riemann[11], mathematicians throughout the world have been unable to prove or disprove the validity of this hypothesis.

Many important examples of Dirichlet series are obtained from the ζ-function, and reflect various arithmetic properties of the integers, for example:

(1.3.5)　$\zeta^{-1}(z) = \Sigma \mu(n)/n^z$, where μ is the Möebius function, i.e.,

$\mu(1) = 1$

$\mu(n) = 0$ if n is not square-free

$\quad\quad\quad = (-1)^k$ if n is the product of k distinct primes;

(1.3.6)　$\zeta^2(z) = \Sigma d(n)/n^z$, where d is the divisor function, i.e. $d(n)$ is the number of positive divisors of n ;

4

$$(1.3.7) \qquad \zeta'(z) \;=\; \Sigma - \log(n)/n^z \qquad (cf.(1.1));$$

$$(1.3.8) \qquad \zeta'(z)/\zeta(z) \;=\; \Sigma \, \Lambda(n)/n^z, \quad \text{where}$$

$$\Lambda(n) \;=\; \log(n) \quad \text{if} \;\; n \;\; \text{is a power of the prime} \;\; p,$$

$$=\; 0 \quad \text{otherwise};$$

$$(1.3.9) \qquad \zeta(z-1)/\zeta(z) \;=\; \Sigma \, \phi(n)/n^z, \quad \text{where} \;\; \phi \;\; \text{is the Euler function.}$$

In all the above examples[12] except (1.3.9), the Dirichlet series are absolutely convergent in H_1. The Dirichlet series in (1.3.9) is absolutely convergent in H_2.

2 Classical number theory

Let Z denote the integers, and Q the rationals (its field of fractions). If K is a finite algebraic extension of Q, then the integral closure[13] of Z in K, denoted by A, is called the <u>ring of integers</u> of K. Such rings have many properties in common with the ring Z, as we now indicate. Various proofs and definitions have been omitted, but will be given in more generality

We start by remarking that if $r = \dim K/Q$, then A is a free abelian group of rank r.

2.1   UNITS

Let U denote the group of units in A. The subgroup of elements of finite order is clearly $\mu(K)$, the group of roots of unity in K, which is a finite cyclic group. The quotient $U/\mu(K)$ is a free abelian group. Now there are precisely r isomorphisms of K into C, of which say r_1 map into R, and the remainder occur in r_2 conjugate pairs, so that $r = r_1 + 2r_2$. The rank of $U/\mu(K)$ is $r_1 + r_2 - 1$. Putting these results together we have <u>Dirichlet's Unit Theorem</u>.

<u>Theorem 2.1.1</u>

$$U \cong Z^{r_1 + r_2 - 1} + \mu(K)$$

A unit which forms part of a basis for the free part of U is called a <u>fundamental unit</u>.

In the case that $K = Q(\sqrt{d})$, where $d \in Z$ is square-free, the group U is easy to describe. If the group of roots of unity in $Q(\sqrt{d})$ has order m, then $Q(\sqrt{d})$ contains a subfield isomorphic with $Q(\omega)$, where $\omega = \exp(2\pi i/m)$

6

Since $\deg Q(\omega)/Q = \phi(m)$, where ϕ is the Euler function, and $\deg Q(\sqrt{d})/Q = 2$, we must have $\phi(m) = 1$ or 2, which can only happen if $m = 1,2,3,4$ or 6. Since $-1 \in U$, and $(-1)^2 = 1$, m is necessarily even. In fact one can show[14] that $m = 2$ except when $d = -1$ $(m = 4)$ or $d = -3$ $(m = 6)$.

If $d > 0$, then $r_1 = 2$, $r_2 = 0$ so U has rank 1. If $d < 0$, then $r_1 = 0$, $r_2 = 1$ so U is finite and cyclic.

<u>Example 2.1.2</u>   If $K = Q(\sqrt{2})$, then $\mu(K) = \{-1,1\}$ so $U \cong Z \oplus Z/2Z$.

A fundamental unit is $1 + \sqrt{2}$, so any unit is uniquely expressible in the form $\pm(1 + \sqrt{2})^n$ for some $n \in Z$.

<u>Example 2.1.3</u>   If $K = Q(\sqrt{94})$, the only fundamental units are the numbers $\pm 2143295 \pm 221064\sqrt{94}$.

There is no formula for the fundamental units even in the quadratic case. They may be obtained by means of continued fractions[15].

<u>Example 2.1.4</u>   For p a prime integer, let $\omega = \exp(2\pi i/p)$, and $K = Q(\omega)$. The ring of integers A is $Z[\omega]$ and $r_1 = 0$, $r_2 = (p-1)/2$ for $p \geq 5$. The element $\xi_t = \omega^t + \omega^{-t} = 2\cos(2\pi t/p)$ for $t \in Z$ is a unit. In general a fundamental set of units is not known.

2.2   IDEAL THEORY

The ring A is Noetherian, and every non-zero prime ideal is maximal. Since A is by definition integrally closed, A is[16] a <u>Dedekind domain</u>, and so every non-zero ideal is uniquely expressible as a product of prime ideals. This is the analogue of unique factorisation. However, in general not every ideal is principal, and the <u>ideal class group</u> measures this deficiency. This group is trivial if and only if A is a principal ideal domain. In general (and this is an important result) this group is finite. Its order is usually denoted by $h = h(K)$, and is called the <u>class-number</u> of K.

If I is a non-zero ideal of A, then the quotient ring A/I is finite.
Let $N(I)$ denote its order, which is called the <u>norm</u> of I. The norm is
multiplicative, i.e. $N(IJ) = N(I)N(J)$. Moreover, for any positive integer
n, the set $\{I \lhd A: N(I) = n\}$ is finite.

2.3   THE ZETA FUNCTION OF K

Let $\zeta(z,K)$ denote the Dirichlet series

$$(2.3.1) \qquad \sum_{\substack{I \lhd A \\ I \neq 0}} 1/N(I)^z \;=\; \sum_{n} \Big(\sum_{\substack{I \lhd A \\ N(I)=n}} 1/n^z \Big) .$$

By analogy with the classical case, there is the following proposition[17] :

<u>**Proposition 2.3.2**</u>   *The Dirichlet series $\zeta(z,K)$ converges absolutely in H_1,
and extends to a meromorphic function on the complex plane with a simple pole
at $z = 1$.   The infinite product*

$$(2.3.2) \qquad \prod_{P \in \max(A)} (1 - N(P)^{-z})^{-1} \;=\; \prod_{n} \Big(\prod_{\substack{p \,\in\, \max(A) \\ N(P) \,=\, n}} (1 - n^{-z})^{-1} \Big)$$

*converges in H_1 to $\zeta(z,K)$.   Moreover, this infinite product converges
absolutely in H_1, uniformly on compact subsets.*

As in 1.3.2 the proof of 2.3.2 depends essentially on unique factorisation.

<u>Remark 2.3.3</u>   If P is a maximal ideal of A, then $N(P)$ is a power of the
prime p, where p generates the ideal $Z \cap P$ in Z.   Thus the infinite
product (2.3.2.1) is indexed by prime powers.

The following theorem[18] relates the behaviour of $\zeta(z,K)$ near its pole
to the field K.

__Theorem 2.3.4__   *The residue of* $\zeta(z,K)$ *at the pole* $z = 1$ *is*

$$(2.3.4.1) \qquad \frac{2^{r_1}(2\pi)^{r_2} hR}{|\mu(K)|\sqrt{D}}$$

where h *is the class number,* r_1 *and* r_2 *are as defined in 2.1,* D *is the discriminant and* R *the regulator of* K.

This theorem and the two previous propositions indicate how the analytic properties of $\zeta(z,K)$ capture algebraic properties of K and its ring of integers. Other Dirichlet series and meromorphic functions may be obtained from $\zeta(z,K)$ as in §1.3.

2.4   THE FUNCTIONAL EQUATION

It may be shown that the ζ-function of K satisfies[19] the functional equation (2.4.1):

$$\zeta(z,K) = |D|^{\frac{1}{2}-z}\left(\frac{\pi^{z-1/2}\Gamma(1/2 - z/2)}{\Gamma(z/2)}\right)^{r_1}\left(\frac{(2\pi)^{2z-1}\Gamma(1-z)}{\Gamma(z)}\right)^{r_2}\zeta(1-z,K) \ .$$

It follows from 2.3.2 as for $\zeta(z)$ that $\zeta(z,K)$ does not vanish for $z \in H_1$. Since $\Gamma(z)$ does not vanish anywhere, has simple poles at the non-positive integers but no other poles, we deduce:

__Lemma 2.4.2__   *The zeros of* $\zeta(z,K)$ *in the region* $\{z: \mathrm{Re}(z) < 0\}$ *can be described as follows:*

(i) *if* $r_2 \neq 0$ *then for any* $m \in N$, $z = -2m$ *is a zero with multiplicity* $r_1 + r_2$ *and* $z = 1 - 2m$ *is a zero with multiplicity* r_2;

(ii) *if* $r_2 = 0$, *then for any* $m \in N$, $z = -2m$ *is a zero with multiplicity* r_1;

(iii) *there are no other zeros in this region.*

__Corollary 2.4.3__   *The zeta function of* K *determines* $r_1, r_2, \dim K/Q$ *and* $|D|$.

__Proof.__   From 2.4.2 we see that r_1 and r_2 are determined by the multiplicities of the zeros of the zeta function in the region $\{z: \mathrm{Re}(z) < 0\}$, and so $\dim K/Q = r_1 + 2r_2$ is determined.   Substituting these values in equation (2.4.1) enables $|D|$ to be determined.

__Corollary 2.4.4__   *For quadratic fields* K, *the zeta function determines* K.

__Proof.__   Suppose $K = Q(\sqrt{d})$ where d is a square free integer.   Then[20] we have $D = d$ if $d \equiv 1 \bmod 4$, but $D = 4d$ if $d \not\equiv 1 \bmod 4$, and so

$$d = (-1)^{1+r_1}|D| \qquad \text{if} \quad |D| \equiv 1 \bmod 4$$

$$= (-1)^{1+r_1}|D|/4 \quad \text{if} \quad |D| \not\equiv 1 \bmod 4 \ .$$

3 Artin's thesis

E. Artin, in his thesis[21] in 1921, observing that the integral domain $F_p[X]$ of polynomials over the finite field F_p with p elements (p is prime) has much in common with the ring Z of integers, examined the analogues of the quadratic fields. That is to say, let $f(X) \in F_p[X]$ be a polynomial of degree at least 1 and which is square-free, and let K be the splitting field of the polynomial $Y^2 - f(X)$ (in the variable Y) over the field $F_p(X)$ of rational functions of X (which is the field of fractions of $F_p[X]$). Let A be the integral closure of $F_p[X]$ in K. Artin investigated the properties of the ring A by analogy with the classical case.

If we write $Y = \sqrt{f}$, then any element of K can be expressed uniquely in the form $a + b\sqrt{f}$ where $a, b \in F_p(X)$. If p is odd, then $a + b\sqrt{f} \in A$ if and only if $a, b \in F_p[X]$. (This is similar with the classical case, where the integers of $Q(\sqrt{d})$ are of the form $1/2(a + b\sqrt{d})$, where a, b are rational integers. Since $p \neq 2$, we can divide by 2 in $F_p[X]$.) Thus if $p \neq 2$ (which we tacitly assume throughout this number) $A = \{a + b\sqrt{f} \in K : a, b \in F_p[X]\}$. Clearly A is a free $F_p[X]$-module of rank 2.

3.1   THE UNITS OF A

The theory of units in the classical case depends on whether or not $\sqrt{d}$ is real. Artin introduces the analogue of this as follows. Let $F_p(X)_\infty$ be the field of finite Laurent series in X^{-1}. An element of this field is a formal sum of the form $\sum_{n=-\infty}^{n=\infty} a_n X^n$ such that for some $m \in Z$, $a_n = 0$ if $n \geq m$. This field contains $F_p[X]$, and hence $F_p(X)$ as a subfield.

<u>Definition 3.1.1</u>   We say that $\sqrt{f}$ is <u>real</u> if $Y^2 - f$ factorises over $F_p(X)_\infty$, i.e. if there exists a $g \in F_p(X)_\infty$ such that $g^2 = f$.   We say that $\sqrt{f}$ is <u>imaginary</u> if it is not real.

Thus $\sqrt{f}$ is real if and only if there is an embedding of K into $F_p(X)_\infty$ which is the identity on $F_p(X)$.

<u>Lemma 3.1.2</u>   *If* $f(X) = a_n X^n + \ldots + a_0$, *where* $a_n \neq 0$, *then* $\sqrt{f}$ *is real if and only if*

    (i) $\deg(f) = n$ *is even,*

    (ii)   *the leading coefficient* a_n *has a square root in* F_p.

<u>Proof</u>.   If $\sqrt{f}$ is real, clearly (i) and (ii) are satisfied.   Conversely, if (i) and (ii) are satisfied we can write $f = a^2 X^{2m}(1+h)$, where $a^2 = a_n$, $2m = n$ and h is a polynomial in X^{-1} of degree $2m$ with $h(0) = 0$. Since[22] the binomial coefficient $\binom{1/2}{r}$ is a rational number of the form $a/2^m$ with $a \in \mathbb{Z}$ it can be considered as an element of F_p.   The power series $\sum_{r=1}^{\infty} \binom{1/2}{r} h^r$ converges in $F_p(X)_\infty$ to a square root of $1+h$, and so

$$g = a X^m \sum_{r=1}^{\infty} \binom{1/2}{r} h^r$$

is an element of $F_p(X)_\infty$ whose square is f.

Let U denote the group of units of A.   The subgroup of elements of finite order is $\mu(K) = \mu(F_p) = F_p^*$ which is a finite cyclic group of order $p-1$.   Artin shows[23] that U/F_p^* is zero if $\sqrt{f}$ is imaginary, but is infinite cyclic if $\sqrt{f}$ is real, in the second case using continued fractions as in the classical case to obtain fundamental units.

3.2   IDEAL THEORY

The ring A is Noetherian, every non-zero prime ideal is maximal and A

12

is integrally-closed, so again A is a Dedekind domain, and hence every non-zero ideal is a unique product of prime ideals.   The ideal class group is again finite[24]; its order is as usual denoted by h.

If I is a non-zero ideal of A, then A/I is finite; let N(I) be its order.   For any $n \in N$ the set $\{I \lhd A: N(I) = n\}$ is finite.   We define the ζ-function of the extension $K/F_p(X)$ by the Dirichlet series

$$(3.2.1) \qquad \zeta(z,K/F_p(X)) \;=\; \sum_{\substack{I \lhd A \\ I \neq 0}} 1/N(I)^z.$$

Because of unique factorisation of ideals and the fact that N is multiplicative, we could equally well define the ζ-function as the infinite product

$$(3.2.2) \qquad \zeta(z,K/F_p(X)) \;=\; \prod_{P \in \max(A)} (1 - N(P)^{-z})^{-1}.$$

If I is a non-zero ideal of A, then A/I is a finite vector space over F_p, say of dimension deg(I), so that $N(I) = p^{\deg(I)}$.   We define the Z-function of $K/F_p(X)$ to be the power series/infinite product

$$(3.2.3) \quad Z(t,K/F_p(X)) \;=\; \sum_{\substack{I \lhd A \\ I \neq 0}} t^{\deg(I)} \;=\; \prod_{P \in \max(A)} (1 - t^{\deg(P)})^{-1}.$$

The relation between the ζ-function and the Z-function is that

$$\zeta(z,K/F_p(X)) = Z(p^{-z},K/F_p(X)) \quad \text{and} \quad Z(t,K/F_p(X)) = \zeta(-\log_p t,K/F_p(X)).$$

<u>Proposition 3.2.4</u>   *The ζ-function of $K/F_p(X)$ is absolutely convergent in H_1.*

<u>Proof.</u>   Consider the infinite product representation (3.2.2) of $\zeta(z,K/F_p(X))$. For a fixed value of z, this product is absolutely convergent if[25] (and only if) the series $\sum_{P \in \max(A)} 1/N(P)^z$ is absolutely convergent.   If $P \in \max(A)$ then the ideal $P' = P \cap F_p[X]$ of $F_p[X]$ is prime, and is non-zero, so is maximal.   Since A/P is an extension of $F_p[X]/P'$, we know that $N(P) \geq p^r$

where r is the degree of any polynomial which generates P'. Conversely, if P_1 is a maximal ideal of $F_p[X]$, there are[26] at most 2 maximal ideals P' of A such that $P' \cap F_p[X] = P_1$. Hence we have

$$\sum_{P \in \max(A)} \left|1/N(P)^z\right| \leq 2 \sum_{k=1}^{\infty} \sum_{\substack{f \in F_p[X] \\ f \text{ monic} \\ \text{degree } k}} \left|1/p^{kz}\right| \leq 2 \sum_{k=1}^{\infty} p^k / p^{kz}$$

and the series on the right of these inequalities, being a geometric series with ratio less than 1, is convergent.

<u>Corollary</u> (3.2.5)   *The Z-function of* $K/F_p(X)$ *is absolutely convergent in the disc* $\{t \in C : |t| < p^{-1}\}$.

Artin next showed that $Z(t, K/F_p(X)) = P(t, K/F_p(X))/(1 - pt)$, where $P(t, K/F_p(X))$ is a polynomial in t, of degree less than the degree of f. In fact he gave an explicit value of its degree, depending on whether or not $\sqrt{f}$ is real and whether n is even[27] and these results can be extracted from (3.3.4) and (3.3.7) below.

Thus Z and ζ extend to meromorphic functions on the complex plane. The Z-function has a single simple pole at $t = 1/p$, and so the ζ-function (being periodic with period $2\pi i/p$) has simple poles at $z = 1 + 2k\pi i/p$ for any $k \in Z$. The residues at these poles involve[28] the class number of K, and can be determined from (3.3.9).

3.3   THE FUNCTIONAL EQUATION

Suppose that $f \in F_p[X]$ is square-free of degree n. The Z-function satisfies the functional equation[29]

$$(3.3.1) \quad Z(1/pt, K/F_p(X)) = \left(\frac{1 - pt}{1 - 1/t}\right)\left(\frac{1}{pt^2}\right)^g Z(t, K/F_p(X)) \quad \text{if} \quad \sqrt{f} \text{ is}$$

imaginary and n is odd,

14

(3.3.2) $Z(1/pt,K/F_p(X)) = \left(\dfrac{1 - p^2t^2}{1 - 1/t^2}\right)\left(\dfrac{1}{pt^2}\right)^g Z(t,K/F_p(X))$ if f is imaginary,

n even,

(3.3.3) $Z(1/pt,K/F_p(X)) = \left(\dfrac{1 - pt}{1 - 1/t}\right)^2 \left(\dfrac{1}{pt^2}\right)^g Z(t,K/F_p(X))$ if $\sqrt{f}$ is real,

where g is the genus[30] of $K/F_p(X)$, which is[31] $(n-1)/2$ if n is odd,

but $n/2-1$ if n is even.   If we define the Z-function of K, $Z(t,K)$ to

be

(3.3.4) $Z(t,K/F_p(X))/(1-t)$ if $\sqrt{f}$ is imaginary, n odd,

$Z(t,K/F_p(X))/(1-t^2)$ if $\sqrt{f}$ is imaginary, n even,

$Z(t,K/F_p(X))/(1-t)^2$ if $\sqrt{f}$ is real.

then the above functional equations (3.3.1-3) can be condensed into the single

equation

(3.3.5) $Z(1/pt,K) = (pt^2)^{1-g} Z(t,K)$.

Remark 3.3.6   We shall see in Chapter 5 that, as the notation suggests,

$Z(t,K/F_p(X))$ depends on the extension $K/F_p(X)$, and thus on A, whereas

$Z(t,K)$ depends only on the field K.   This situation does not arise in the

classical case.

As we have said, Artin showed[32] essentially that

(3.3.7) $Z(t,K) = P(t,K)/(1 - pt)(1 - t)$ where $P(t,K)$ is a polynomial of

degree 2g, and satisfies the functional equation

(3.3.8) $P(1/pt,K) = (pt^2)^{-g}P(t,K)$.

Artin's results about the residues of the ζ-functions are summarised in

the following;

Theorem 3.3.9   *The residue of* $\zeta(z,K)$ *at* $z = 1$ *is* $\dfrac{hp^{1-g}}{(p-1)\log p}$.

This result is equivalent to either of the statements (i) the residue of

$Z(t,K)$ at $t = 1$ is $h/p-1$ or (ii) $P(1) = h$.

3.4 THE RIEMANN HYPOTHESIS

Artin observed, by making specific calculations in many cases that the non-integral zeros of the ζ-function appeared to lie on the line $Re(z) = 1/2$. He conjectured (by virtue of the relation between the ζ-function and the Z-function) the following.

<u>Riemann Hypothesis for quadratic function fields</u>. (3.4.1) *The zeros of the* Z-*function of a quadratic extension of* $F_p(X)$ *are either trivial* (*i.e.* ± 1) *or lie on the circle* $|t| = 1/\sqrt{p}$.

It follows from (3.3.8) that $P(t,K) = \prod\limits_{i=1}^{2g} (1 - \alpha_i(t))$ where $\alpha_i \alpha_{g+i} = p$. The Riemann Hypothesis is that $|\alpha_i|^2 = p$.

3.5 THE POINTS AT INFINITY

The ideas of Artin extend to the study of the integral closure of $F_p[X]$ in some finite separable extension K of $F_p(X)$. One can define exactly as in (3.2.2) and (3.2.3) the corresponding ζ- and Z-functions, and as the following questions.

(3.5.1) Are the ζ- and Z-functions absolutely convergent at any points, i.e. are they holomorphic in some domain?

(3.5.2) Do they extend to meromorphic functions on the complex plane, and if so where are the zeros and poles and what are the residues?

(3.5.3) Do they satisfy a functional equation?

(3.5.4) Is the Z-function rational?

(3.5.5) Is the generalised Riemann Hypothesis true, i.e. do the zeros of the Z-function have absolute value $1/\sqrt{p}$?

We have seen that even in the quadratic case that there seem to be three separate cases to consider (depending on whether $\sqrt{f}$ is real or imaginary).

But these distinctions are somewhat artificial as the next example shows.

<u>Example 3.5.6</u>   Let K be the field obtained by adjoining to $F_7(X)$ the square roots of $f(X) = 3X^4 + 1$.   We shall see later[33] that

$$Z(t,K) = (1+7t^2)/(1-t)(1-7t).$$ Now since $Y = \sqrt{f}$ is imaginary, we have

$$Z(t,K/F_7(X)) = (1+t)(1+7t^2).$$

Now consider the elements $X_1 = 1/X$ and $Y_1 = Y/X^2$ of K.   The field K is obtained by adjoining to $F_7(X_1)$ the element Y_1 which is a square root of $X_1^4 + 3$, so that Y_1 is real, and so

$$Z(t,K/F_7(X_1)) = (1-t)(1+7t^2)/(1-7t).$$

If we put $X_2 = 1/X-2$ and $Y_2 = Y/(X-2)^2$, then K is also obtained from $F_7(X_2)$ by adjoining Y_2 which is a square root of $5X_2^3 + 2X_2^2 + 3X_2 + 3$. Again Y_2 is imaginary, but Y_2^2 has odd degree in X_2, so

$$Z(t,K/F_7(X_2)) = (1 + 7t^2)/(1-7t).$$

This example shows that the same field can occur as a real or imaginary quadratic extension of either type depending on the choice of subfield of rational functions.

Moreover, since the three Z-functions are different, the three corresponding rings of integers, the integral closures of $F_7[X]$, $F_7[X_1]$ and $F_7[X_2]$ are not isomorphic.

If K is a finite algebraic extension of $F_p(X)$, then K is a finitely generated extension of a finite field of transcendence degree[34] one. Conversely[35] any finitely generated extension of a finite field of transcendence degree one is a finite separable extension of $F_p(X)$.

<u>Definition 3.5.7</u>   A <u>function field in</u> n <u>variables over a field</u> F is a finitely generated extension of F of transcendence degree n.   A function field in one variable over a finite field is called an <u>algebraic function field</u>.   If K is an algebraic function field, its <u>field of constants</u> (or

constant field) is the algebraic closure of the prime field in K.   A <u>global</u> <u>field</u> (or A-<u>field</u>) is either an algebraic number field, or an algebraic function field.

We shall see in the next section that there is a close analogy between algebraic number fields and algebraic function fields, which justifies the introduction of a common label.

Suppose that K is an algebraic function field, with field of constants k.   A choice of embedding $k(X) \to K$ (or equivalently a choice of element $x \in K$ transcendental over k) is called a model for K.   Each model has an associated ring of integers, which as example 3.5.6 shows depends on the model. The explanation of this and the relation between various models is geometrical, and will be given in Chapter 9.   We give a simple sketch now.

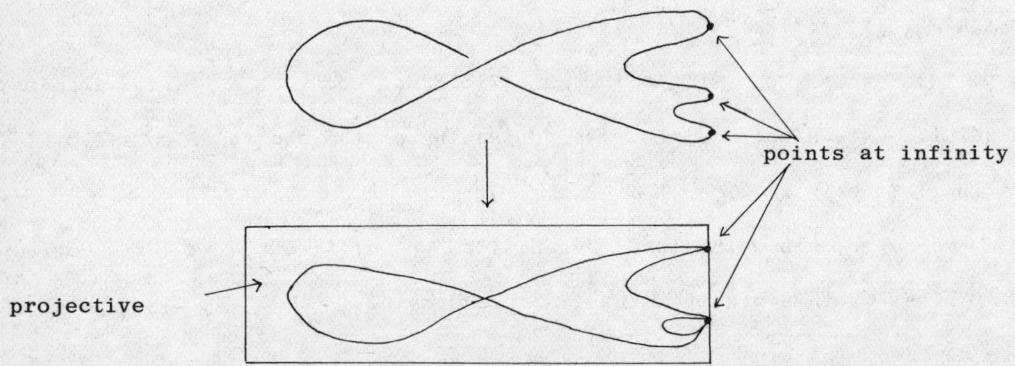

An algebraic function field K corresponds to a 'smooth' curve Γ in some projective space, and each model corresponds to a projection from Γ on to a projective line.   The ring of integers of a model corresponds to that part of Γ which projects onto the affine line, which consists of all but a finite set of points of Γ.   In other words, the model 'ignores information at infinity'.   However, since there is only a finite set of points at infinity, not too much is lost.   The equations (3.3.4) are essentially putting back the points at infinity.

Although this situation does not arise in the classical case (for there is a unique embedding of Q into any algebraic number field) it can still be useful to consider points at infinity.   In order to do this, one introduces valuations.

4 Valuation theory

In this chapter we summarise the essential properties of valuations.   These
are then used in Chapter 5 to provide a unified theory for both algebraic
number fields and algebraic function fields.

4.1   BASIC DEFINITIONS

<u>Definition 4.1.1</u>   A <u>valuation</u> on a field K is a function $v: K \to R$ such
that

(4.1.1.1) $v(x) \geq 0$ for all $x \in K$, and $v(x) = 0$ if and only if $x = 0$,

(4.1.1.2) $v(x + y) \leq v(x) + v(y)$ $\forall\, x,\, y \in K$,

(4.1.1.3) $v(xy) = v(x)v(y)$ $\forall\, x,y \in K$.

The subgroup $v(K^*) \subset R^*$ is called the <u>value group</u> of the valuation v,
and if it is a discrete group, then v is called a <u>discrete</u> valuation.

It follows from the definition that v is a metric on K, and that with
the induced topology K is a topological field.   If v

(4.1.1.4) $v(x + y) \leq \max\{v(x),v(y)\}$ $\forall\, x,y \in K$,

then v is said to be <u>non-archimedean</u>, or <u>ultrametric</u>.   If it is not non-
archimedean, then it is called <u>archimedean</u>.   It is easy to prove[36]

<u>Lemma 4.1.2</u>   *If* v *is an ultrametric valuation on the field* K *and*
$x_1,\ldots,x_n \in K$ *are such that* $v(x_1) > v(x_i)$ *for* $i = 2,\ldots,n$ *then*
$v(x_1 + \ldots + x_n) = v(x_1)$.

A valuation $v: K \to R$ is <u>complete</u> if every Cauchy sequence
converges.   One can associate in the usual way[37] to any valuation v on a

field K its completion $i_v : K \to K_v$ which is characterised by the properties

that K_v is complete, and that the image of i_v is dense in K_v. Of course,

K_v is itself a field.

Suppose that v,w are two valuations on the field K. The following

conditions are equivalent[38]::

(4.1.3.1) the topologies induced by v and w coincide,

(4.1.3.2) $v(x) < 1$ if and only if $w(x) < 1$,

(4.1.3.3) $\exists \, r \in R$ such that $v(x)^r = w(x)$, $\forall x \in K$

(4.1.3.4) there is a topological isomorphism $\phi : K_v \to K_w$ such that

$\phi i_v = i_w$.

<u>Definition 4.1.4</u> Two valuations v,w on K which satisfy the conditions

(4.1.3.1-4) are said to be <u>equivalent</u>. An equivalence class of valuations on

K is called a <u>spot</u> on K.

Various properties of valuations which are preserved by equivalence can

thus be attributed to the spots they determine. For example, a spot S is

non-archimedean if some (and hence all) $v \in S$ is non-archidedean. For most

purposes, the notions of 'valuation' and 'spot' can be identified, and we

write K_S for K_v when S is the spot determined by v.

<u>Example 4.1.5</u> For any field K the spot defined by the valuation $v(x) = 1$

if $x \neq 0$, $v(0) = 0$ is called the <u>trivial</u> spot. It induces on K the

discrete topology, and is a complete and discrete spot. If K is finite,

the trivial spot is the only one.

<u>Example 4.1.6</u> The spot defined by 'absolute value' on Q is called the

<u>infinite spot</u>, and is denoted by ∞. It is not discrete, but it is

archimedean. Its completion is $Q \hookrightarrow R$.

<u>Example 4.1.7</u> Let A be a Dedekind domain, and K its field of fractions.

Let P be a maximal ideal of A. For any $a \in A$, define $\mathrm{ord}_P(a) = $ largest n such that $a \in P^n$ (if $a \neq 0$), and $\mathrm{ord}_P(0) = 0$.

Hence if $a \in R$ is non-zero then $\displaystyle\prod_{P \in \max(A)} P^{\mathrm{ord}_P(a)}$ is the unique

factorisation of the ideal generated by a.   For $x \in K$,   define

$\mathrm{ord}_P(x) = \mathrm{ord}_P(a) - \mathrm{ord}_P(b)$ if $x = a/b$.   Choose some constant $c > 1$, and

let $v_P(x) = c^{\mathrm{ord}_P(x)}$.   The function v_P is a valuation, called a P-<u>adic</u>

<u>valuation</u> and the spot it determines (which is independent of the choice of c)

is called the P-<u>adic spot</u>, and is denoted by P.   If P and P' are distinct

maximal ideals of A, then clearly the P-adic and P'-adic spots are distinct.

For example the spot defined on the field of rational functions F(X)

over the field F by the valuation

$$v(f/g) = c^{\deg(f) - \deg(g)} \quad \text{where } c > 1, \; f,g \in F[X]$$

is independent of the choice of the constant c, and is called the <u>infinite</u>

<u>spot</u>, denoted by ∞.   We observe that this valuation is the composite of the

X -adic valuation on F(X) with the automorphism of F(X) which takes X

to 1/X.

The following two propositions show an analogy between the fields Q and

F(X).

<u>Proposition 4.1.8</u>   *The non-trivial spots on· Q are*

    (1)   *the infinite spot,*

    (2)   *the P-adic spots, where P is a maximal ideal in Z .*

All these spots are distinct, and all but the infinite spot are non-archimedean.   All of the completions are locally compact.

The proof of 4.1.8 is essentially the same as for 4.1.9 below.   We note

that if k is a subfield of K, then any spot on K restricts to a spot on

k.

<u>Proposition 4.1.9</u> *Let* **F** *be a field.   The non-trivial spots on* **F(X)**

which are trivial on **F** *(i.e. all the non-trivial spots if* **F** *is finite) are*

  (1) *the infinite spot*

  (2) *the* **P**-*adic spots, where* **P** *is a maximal ideal in* **F[X]**.

All these spots are distinct, and all are non-archimedean.   If **F** *is finite,*

then all the completions are locally compact.

<u>Proof</u>   First we note that any spot on F(X) is non-archimedean since[39] its

restriction to F is non-archimedean.   Let S be a spot on F(X), and

$v \in S$ a representative valuation.

If $v(X) > 1$, then $v(X^n) = v(X)^n > v(X)$, and so $v(f) = v(X)^d$ if

$f \in F[X]$ is a polynomial of degree d (using 4.1.2) and hence it follows

that S is the infinite spot.

If $v(X) \leq 1$, then $v(f) \leq 1$ for any polynomial $f \in F[X]$.   Since S

is non-trivial, there is a $g \in F[X]$ such that $v(g) < 1$, and without loss

of generality we can assume that g is irreducible.   If P denotes the

prime ideal $\langle g \rangle$ generated by g, then S is the P-adic spot, as we now

prove.

If $h \in F[X]$ and $g \nmid h$, then there exist $a, b \in F[X]$ such that

$ag + bh = 1$, and so $v(ag + bh) = 1$.   Hence $v(h) = 1$, since otherwise

$v(ag + bh) \leq \max(v(ag), v(bh)) < 1$.   Now any element of F[X] may be written

in the form $g^n h/k$ where $h, k \in F[X]$ and $g \nmid h, g \nmid k$, and then

$v(g^n h/k) = v(g)^n v(h)/v(k) = v(g)^n$ so that S is the P-adic spot.

Let $F(X)_P$ denote the completion of F(X) at the P-adic spot, and

$F(X)_\infty$ the completion at the infinite spot, and suppose now that F is finite.

The natural maps $\{\rho_n : F[X] \to F[X]/P^n\}$ are continuous where $F[X]/P^n$ has the

topology induced by the discrete metric, and so the natural map

$\rho : F[X] \to \prod_n F[X]/P^n = B$ is also continuous.   Since B is the product of finite (hence compact) spaces it is compact, so is a complete metric space, and ρ extends (by universality of completion) to an embedding of the closure of F[X], namely $\{a \in F(X)_P : v(a) \leq 1\}$, into B.

The image of ρ ($= \varprojlim F[X]/P^n$) is closed in B, hence compact. The topology on $F(X)_P$ has as basis sets homeomorphic with the set $\{a \in F(X)_P : v(a) < 1\}$, and so is locally-compact.

It follows that $F(X)_\infty$ is locally compact since it is homeomorphic to $F(X)_{(X)}$.

If F is algebraically closed, then the non-trivial spots on F(X) which are trivial on F are thus in 1-1 correspondence with the points of $F \cup \infty = P^1(F)$, the projective line over F. If F is not necessarily algebraically closed but is perfect, there is a modified relationship.

Suppose that P is a maximal ideal in F[X] which is generated by the mimimum polynomial of the element $\alpha \in F^a$; in this case we write $F(X)_\alpha$ for $F(X)_P$. The following is a special case of Hensel's Lemma[40]:

<u>Lemma 4.1.10</u> *Suppose that F is a perfect field, and P is a maximal ideal in F[X] generated by the polynomial f(X). Let F' denote the quotient field F[X]/P, and let $\alpha \in F'$ denote the image of X under the natural map F[X] $\to$ F'. Then we have*

i) the completion of F(X) at the P-adic spot is the composite

$$F(X) \to F'(X) \to F'(X)_\alpha ,$$

i.e. $F(X)_P$ is the field of finite Laurent series F'((X-α)) in (X-α) over the field F' (which is a finite separable extension of F) ,

ii) the algebraic closure of F in $F(X)_P$ is F' .

<u>Proof</u> Since F'(X) is complete, it suffices to show that the image of F(X) in F'(X) is dense, but since F'(X) is dense in F'(X) it suffices to

show that F' is contained in D , the closure of $F[X]$, or (since α generates F' over F) that $\alpha \in D$. To do this, we prove by induction that for each positive integer n there exist $g_n(X) \in F[X]$ and $h_n(X) \in F'[X]$ such that

(4.1.10.1) $g_n(X) - \alpha = (X - \alpha)^i h_n(X)$ where $i \geq n$.

The induction starts with $n = 1$, $g_1(X) = X$, $h_1(X) = 1$. Assume we have inductively constructed g_n , h_n satisfying (4.1.10.1). If $i > n$ then put $g_{n+1} = g_n$ and $h_{n+1} = h_n$. Otherwise, let g_{n+1} be $g_n(X) + g(X)[f(X)]^n$, where $g \in F[X]$ is yet to be determined. Now $g_{n+1}(X) - \alpha = (X - \alpha)^n(h_n(X) + g(X)r(X)^n)$ where $r(X) = f(X)/(X - \alpha)$ is a polynomial in $F'[X]$. Since F is perfect, f has distinct roots and so $r(\alpha) \neq 0$; let $\beta = h_n(\alpha)/r(\alpha)^n$ so that $\beta \in F'$. Now choose $g(X) \in F[X]$ such that $g(\alpha) = \beta$ and $\deg(g) > \deg(h_n)$ so that $h_n(X) + g(X)r(X)^n$ is not identically zero but has α as a root. There is thus a polynomial $h_{n+1}(X) \in F'[X]$ such that $h_n(X) + g(X)r(X)^n = (X - \alpha)^j h_{n+1}(X)$ where $j \geq 1$ and $h_{n+1}(\alpha) \neq 0$, and so

$$g_{n+1}(X) - \alpha = (X - \alpha)^{i+j} h_{n+1}(X) .$$

The proof of 2) is now trivial since the algebraic closure of F' in $F'((t))$ is F' .

<u>Corollary 4.1.11</u>   *If F is a perfect field, then the non-trivial spots on $F(X)$ which are trivial on F are in $1-1$ correspondence with the points of $P^1(F^a)/G$ where F^a/F is the algebraic closure of F , and G is the Galois group of F^a/F .*

This result is analogous to the fact that there is a $1-1$ correspondence between the maximal ideals of $F[X]$ and the points of F^a/G . We shall pursue this later, in Chapter 6.

25

4.2   EXTENSION OF VALUATIONS AND SPOTS

For any field F, let $P(F)$ denote the set of non-trivial spots on F.
Suppose that $S \in P(F)$ and that v is a valuation representing S. If E/F
is an extension and w is a valuation on E representing a spot T on E
whose restriction to F is v, then we say that w __extends__ v or that T
__extends__ S, and write $T|S$. If σ is an automorphism of the field E over
F then $w\sigma: E \to R$ is also a valuation which extends v, so that if E/F is
a normal extension $\mathrm{Gal}(E/F)$ acts on the set of spots which extend S.
There are three important cases when one can describe precisely all such
extensions.

(4.2.1)   __Case 1__. F *is complete, and* E/F *is a finite extension.* Since
any two norms on a finite dimensional vector space over a complete field are
equivalent, there is at most one spot extending S. There is such an extension[41] determined by the valuation

$$v(e) = v(N_{E/F}(e))^{1/n} \quad \text{for all } e \in E$$

where $n = \deg(E/F)$ and $N_{E/F}: E \to F$ is the norm function.

    Clearly E is then complete, and is archimedean or non-archimedean
according to whether F is. Since E as a topological vector space over F
is isomorphic with F^n with norm $|(x_1,\ldots,x_n)| = \max_i v(x_i)$ it follows that
$B_E \cong B_F^n$ (where B_E, B_F are the closed unit balls in E and F) so that E
is locally compact if and only if F is locally compact.

(4.2.2)   __Case 2__. F *is complete, and* F^a/F *is the algebraic closure.* By
case 1 there is a unique spot on F^a extending S which is represented by
the valuation

$$v(e) = \left[v(N_{F(e)/F}(e)) \right]^{1/\deg(F(e)/F)} \quad \text{for } e \in F.$$

If F^a/F is an infinite extension then F^a may not be complete[42] but its

26

completion is algebraically closed[43].

(4.2.3)   <u>Case 3</u> F *is not necessarily complete, but* E/F *is a finite separable extension.*

In this case, if $n = \deg(E/F)$ then there is at least one spot and at most n spots which extend S .

Let $\alpha \in E$ be a primitive element, and let $f(X) \in F[X]$ be its minimum polynomial. Let B be the splitting field of f over F_v , which by case 1 has a unique spot extending S . Let $K \subset B$ be the subfield generated over F by the roots of f, i.e. K is the splitting field of f over F and hence the normal closure of E . By abuse of notation, let v denote the valuations on B and K obtained by first extending y on F_v to B , and then restricting to K . Since E/F is separable, there are precisely n

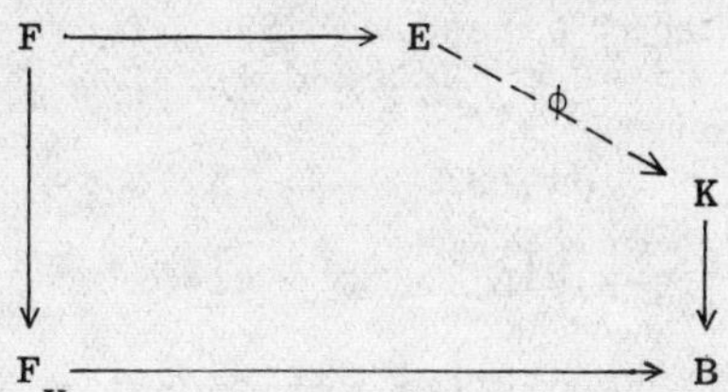

embeddings $\phi : E/F \to K/F$ and each of these induces a valuation v_ϕ on E by the formula $v_\phi(e) = v(\phi(e))$ for all $e \in E$, with a corresponding spot S_ϕ .

Moreover, every spot extending S arises in this way. For if w is a valuation on E extending v , then E_w is generated over F_v by $\alpha \in E_w$ so there is an embedding $j : E_w/F_v \to B/F_v$ which is continuous by uniqueness of valuations on finite extensions of F_v , and since $f(X)$ splits completely in B there is an embedding $\phi : E/F \to K/F$ so that the composite $E \xrightarrow{\phi} K \to B$ is the same as $E \to E_w \xrightarrow{j} B$. Since then $w = v_\phi$, the spot S is equal to S_ϕ .

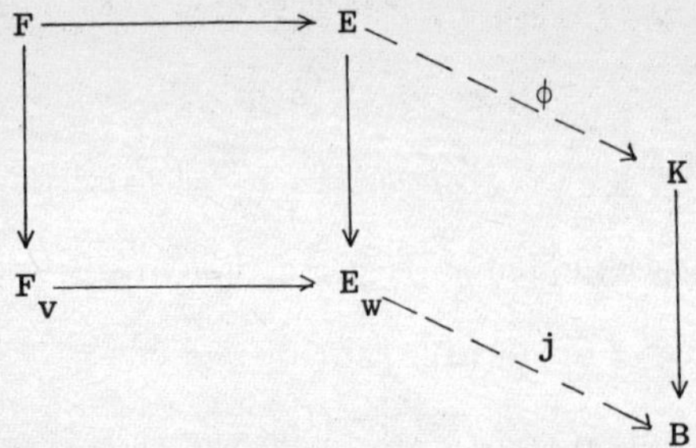

Any two embeddings $E/F \to K/F$ differ by composition with an automorphism of K/F .

<u>Proposition 4.2.4</u> *If K/F is a finite normal extension and S is a spot on F , then the Galois group of K/F acts transitively on the set of spots extending S .*

On the other hand, if two embeddings of E/F in $K/F \subset B/F$ differ by composition with an qutomorphism of B/F_v they determine the same spot since $v(\sigma(b)) = v(b)$ for all $b \in B$, $\sigma \in \mathrm{Gal}(B/F_v)$. The converse is also true:

<u>Lemma 4.2.5</u> *If two embeddings $\phi_1, \phi_2 : E/F \to K/F$ determine the same spot on E , then there is a $\sigma \in \mathrm{Gal}(B/F_v)$ such that $\phi_2 = \sigma\phi_1$.*

<u>Proof</u>. If ϕ_1, ϕ_2 determine equivalent valuations v_1, v_2 then there is a (topological) isomorphism $\psi : E_{v_1}/E \to E_{v_2}/E$ such that $\psi_2 = \psi\phi_1$. Restricted to F , ϕ_1 and ϕ_2 are both the natural map $F \to F_v$, so that ψ is the identity on F , hence on F_v by denseness, and so there exists a $\sigma \in \mathrm{Gal}(B/F_v)$ such that $\sigma|E_{v_1} = \psi$.

<u>Corollary 4.2.6</u> *If $f(X)$ is an irreducible separable polynomial over the field F and $E = F[x]/\langle f\rangle$, then there is a $1-1$ correspondence between the set of spots on E which extend a given spot S on F and the set of non-constant irreducible factors of $f(X)$ over F_S .*

<u>Example 4.2.7</u> Consider the extension $Q(\alpha)/Q$ where $\alpha \in R$ satisfies

$\alpha^3 = 2$, and let S be the infinite spot on Q . In this example, using the notation of 4.2.3, $B = C$ and $K = Q(\alpha,\omega)$, where $\omega = \exp(2\pi i/3)$. The three embeddings of $Q(\alpha)/Q$ into $Q(\alpha,\omega)/Q$ are determined by the image of α which is either $\alpha, \alpha\omega$ or $\alpha\omega^2$, the last two being conjugate. The minimum polynomial of α over Q is $X^3 - 2$ which factorises over R into $(X - \alpha)(X^2 + \alpha X + \alpha^2)$, both terms being irreducible. Thus there are two valuations on $Q(\alpha)$ extending the infinite one on Q given by the formulae

$$v_1(a + b\alpha + c\alpha^2) = |a + b\alpha + c\alpha^2|$$
$$v_2(a + b\alpha + c\alpha^2) = |a + b\omega\alpha + c\omega^2\alpha^2|$$

when $a,b,c \in Q$ and where $|.|$ denotes the absolute value in C .

<u>Example 4.2.8</u> Let $f(X) \in F_p[X]$ be a square-free polynomial with leading coefficient a , and let E/F be the splitting field of $Y^2 - f$ over F(X) . This is a separable extension if $p \neq 2$, and is normal. Let v be a valuation on F(X) representing the infinite spot. We consider separately the three cases as in Chapter 3.

(4.2.8.1)   <u>Case 1</u> $\sqrt{f}$ *real*. There are two embeddings of E into $F_p(X)_\infty$ corresponding to the two solutions of $Y^2 = f$ in $F_p(X)$ and so there are two spots extending S , each having completion $F_p(X)_\infty$.

(4.2.8.2)   <u>Case 2</u> $\sqrt{f}$ *imaginary*, deg(f) *even*.

Since a is not a square in F_p , the field $F_p(\sqrt{a})$ is F_q where $q = p^2$. Now $\sqrt{(a^{-1}f)}$ is real, and so there are elements $g_1, g_2 \in F_p(X)_\infty$ such that $g_1^2 = g_2^2 = a^{-1}f$, $g_1 = -g_2$ and so $\sqrt{(ag_1)}$ and $\sqrt{(ag_2)}$ in $F_q(X)_\infty$ are two distinct square roots of f , and these are conjugate under the Galois group (whose non-zero element acts on a Laurent series as the Frobenius map on the coefficients), and so determine same spot, that is the unique spot extending S ; its completion is $F_q(X)_\infty$ in which the algebraic closure of the prime field is F_q of order p^2.

(4.2.8.3) <u>Case 3</u> $\sqrt{f}$ *imaginary*, $\deg(f)$ *odd*.

The splitting field of $Y^2 - f$ over $F_p(X)_\infty$ is $F_p(\sqrt{(aX)})_\infty$ since $F(X) = (aX)X^{2n}(1 + h)$ where $h \in F_p(X)_\infty$ and $v(h) < 1$, and where $\deg(f) = 2n + 1$, and so $\sqrt{f} = \pm\sqrt{(aX)}.X^n \Sigma \binom{1/2}{r}h^r \in F_p(\sqrt{(aX)})_\infty$.

The two corresponding embeddings of E into $F_p(\sqrt{(aX)})_\infty$ are again conjugate under the Galois group (whose non-zero element takes $\sqrt{(aX)}$ to $-\sqrt{(aX)}$, and so determine the same spot, which is the unique spot extending S; its completion is $F_p(\sqrt{(aX)})_\infty$.

Now consider a non-infinite spot S on $F_p(X)$ corresponding to an element $\alpha \in F^a$ (see 4.1.10). The extensions of S to E depend on whether or not α is a root of $f(X)$.

If $f(\alpha) \neq 0$ then $f(X) = f(\alpha)(1 + h)$ where $h(X) \in F_p^a[X]$ has α as a root, and so $\sqrt{f} = \pm \sqrt{(f(\alpha))} \Sigma \binom{1/2}{r}h^r$ is an element of $F_p^a(X)_\alpha$. These two square roots of f lie in $F_p(X)_S$ and so there are two spots extending S.

If $f(\alpha) = 0$, then $f(X) = (X - \alpha)(1 + h)$ where $h \in F_p^a[X]$ and $h(\alpha) = 0$. As before $1 + h$ has a square root in $F_p^a(X)_\alpha$, but $(X - \alpha)$ does not. Thus there is a unique spot extending S obtained from either of the (conjugate) embeddings of $F_p(X)$ into the field obtained by adjoining to $F_p^a(X)_\alpha$ the square roots of $(X - \alpha)$.

<u>Proposition 4.2.9</u> If E/F *is a finite separable extension and* $S \in P(F)$ *then the natural map*

$$F_S \otimes_F E \to \bigoplus_{T \mid S} E_T$$

is an isomorphism of (topological) F_S*-algebras.*

<u>Proof</u> Let $e \in E$ be a primitive element, and let $F(X) \in F[X]$ be its minimum polynomial, so that 'evaluation at e' induces an isomorphism

30

$F[X]/\langle f\rangle \to E$.   Let $m_i(X)$ for $i = 1,\ldots,r$ be the distinct monic irreducible factors of $f(X)$ over F_S, and let T_i denote the spot on E corresponding to m_i (see 4.2.5 and 4.2.6), so that 'evaluation at e' induces an isomorphism $F_S[X]/\langle m_i\rangle \to E_{T_i}$.   The result now follows from the Chinese remainder theorem:

$$
\begin{array}{ccc}
F_S \otimes_F F[X]/\langle f\rangle & \xrightarrow{\ \cong\ } & F_S \otimes_F E \\[4pt]
\Big\downarrow{\scriptstyle\cong} & & \Big\downarrow \\[8pt]
F_S[X]/\langle f\rangle & & \\[4pt]
\Big\downarrow{\scriptstyle\cong} & & \\[8pt]
\displaystyle\bigoplus_{i=1}^{n} F_S[X]/\langle m_i\rangle & \xrightarrow{\ \cong\ } & \displaystyle\bigoplus_{i=1}^{n} E_{T_i}
\end{array}
$$

<u>Corollary 4.2.10</u>   *If* E/F *is a finite separable extension and* $S \in P(F)$ *then*

$$
N_{E/F}(e) = \prod_{T \mid S} N_{E_T/F_S}(e) \quad \text{for all } e \in E.
$$

<u>Proof</u>   We have
$$
\begin{aligned}
N_{E/F}(e) &= \det_F(e\colon E \to E) \\
&= \det_{F_S}(e\colon F_S \otimes_F E \to F_S \otimes_F E) \\
&= \prod_{T \mid S} \det_{F_S}(e\colon E_T \to E_T) \qquad \text{(by 4.2.9)} \\
&= \prod_{T \mid S} N_{E_T/F_S}(e).
\end{aligned}
$$

4.3   VALUATION RINGS

Let F be a field, and S a non-archimedean spot on F represented by the valuation $v \in S$.   The subset

$$
(4.3.1) \qquad R_S = R_S(F) = \{x \in F\colon v(x) \leq 1\}
$$

is a subring of F, and is independent of the choice of $v \in S$, by (4.1.2.2). It is called the <u>valuation ring</u> of F) <u>at</u> S, or the <u>ring of integers at</u>

S , its elements being called __integers at__ S .

__Proposition 4.3.2__   *The ring* R_S *is integrally closed in* $\mathbb{F}$.

__Proof__   Suppose $x \in F$ is integral over R_S and so satisfies an equation

of the form

$$x^n + a_{n-1}x^{n-1} + \ldots + a_0 = 0 \quad \text{with} \quad a_0,\ldots,a_{n-1} \quad \text{in} \quad R_S .$$

If x were not in R_S so that $v(x) > 1$, then we sould have

$v(x^n) > v(a_i x^i)$ for $i = 0,\ldots,n-1$ and hence

$$v(0) = v(x^n + a_{n-1}x^{n-1} + \ldots + a_0) = v(x)^n > 1 \qquad \text{(using 4.1.2)}$$

which is a contradiction.

The ring R_S is a local ring with (unique) maximal ideal

(4.3.3) $m_S = m_S(F) = \{x \in F : v(x) < 1\}$ (the interior of R_S)

and group of units

(4.3.4) $U_S = U_S(F) = \{x \in F : v(x) = 1\}$ (the boundary of R_S)

The quotient field

(4.3.5) $F(S) = R_S/M_S$

is called the __residue field__ of F at S .   The natural map from R_S to $F(S)$

will be denoted by ρ or ρ_S , and is called '__reduction mod S__' .

__Example 4.3.6__   If A is a dedekind domain with field of fractions F and

S is the P-adic spot on F where P is some maximal ideal of A (see

4.1.6) then the ring of integers at S is precisely the localisation A_p of

the ring A at the maximal ideal P , and the residue field is A/P .

If $i : F \to E$ is a field extension and T is a spot on E extending the

spot S on F then by restriction i maps R_S into R_T . Moreover

$i^{-1}(m_T) = m_S$, that is to say i is a __local homomorphism__, and it induces a

(natural) embedding $i : F(S) \to E(T)$.   Clearly we have

deg(E(T)/F(S)) $\leq$ deg (E/F) .   In particular, if S is a non-archimedean spot

on a global field K , then K(S) is finite.

The following lemma is quite straightforward[44].

<u>Lemma 4.3.7</u>   *If* F *is a field, and* S *a non-archimedean spot on* F *with*

completion F_S , *then*

(4.3.7.1)   *the natural embedding* $F(S) \to F_S(S)$ *is an isomorphism,*

(4.3.7.2)   *if* v *is a valuation on* F_S *representing the spot* S *then*

$$\mathrm{Im}(v\colon F_S \to R) = \mathrm{Im}(v\colon F \to R) .$$

Thus the residue field and value group are invariant under completion.
We now consider the effect of algebraic closure.

<u>Lemma 4.3.8</u>   *Let* F *be a field which is complete with respect to the non-*

archimedean spot S . *Let* S^a *denote the unique spot on* F^a *(the algebraic*

closure of F) *which extends* S *(see (4.2.2).   Then* $F^a(S^a)$ *is the*

algebraic closure of F(S) .

<u>Proof</u>   Suppose $\overline{f}(X)$ is a monic polynomial over $F^a(S^a)$.   Let f(X) be a

monic polynomial over R_{S^a} of the same degree such that $\rho f = \overline{f}$.   All the

roots of f lie in F^a , and hence in R_{S^a} by 4.3.2.   Their images in

F(S) are all the roots of $\overline{f}(X)$,   so $F^a(S^a)$ is algebraically closed.

Now suppose that $\overline{x}$ is an element of $F^a(S^a)$ and let $x \in \rho^{-1}(\overline{x}) \subset R_{S^a}$

representative of $\overline{x}$.   Since x is algebraic over F we have f(x) = 0

for some polynomial $f(X) \in F[X]$.   By multiplying f(X) by some element of

F if necessary we can assume that $f(X) \in R_S[X]$ and that 1 is a

coefficient of f(X) .   The reduction $\overline{f}(X)$ of f(X) mod m_S is thus non-

trivial and has $\overline{x}$ as a root, that is to say each element of $F^a(S^a)$ is

algebraic over F(S) .

4.4   INTEGRAL CLOSURE

The relation between integral closure and spots is contained in the following result.

__Proposition 4.4.1__   *If* E/F *is a finite separable extension and* S *a non-archimidean spot on* F, *then the integral closure* A *of* R_S *in E is*

$$\bigcap_{T \mid S} R_T.$$

__Proof__   Since $\bigcap_{T \mid S} R_T$ is integrally closed by 4.3.2 and contains R_S , it contains A. Conversely, supoose that $x \in \bigcap_{T \mid S} R_T$, and let

$m(X) = X^n + a_{n-1} X^{n-1} + \ldots + a_0$ be its minimum polynomial over F. Let B be the splitting field of $m(X)$ over F_S and let R be the ring of integers of B (at the unique spot extending S on F_S). Let $x_1, \ldots, x_n$ be the roots of $m(X)$ in B; if $x_i \notin R$ for some i, then the embedding $F(x) \to B$ which takes x to x_i would define a spot on $F(x)$ which would extend to a spot $T \in S^E$ with the property that $x \notin R_T$. Hence $x_i \in R$ for all i, and since a_j is, up to sign, the j-th elementary symmetric polynomial in the roots we have $a_j \in R$ for all j, and so x is integral over R_S.

4.5 DISCRETE SPOTS

Suppose that S is a discrete spot on the field F represented by the valuation $v \in S$. By definition the value group $v(F^*)$ is a discrete subgroup of R^*, and so is a free abelian group of rank 1 generated by $v(\eta)$ where η is any element of $m_S \backslash m_S^2$. Such an element η is called a __prime element__ of F at S. Any other element $x \in F$ is also a prime element if and only if x $u\eta$ for some $u \in U_S$, that is to say there is a unique prime element up to units.

    For any $r \in Z$, let m_S^r denote the R_S-submodule of F generated by η^r (This notation is consistent since for $r \in N$, the element η^r does generate the ideal m_S^r of R_S).

The function $\mathrm{ord}_S\colon F \to Z$ defined by

(4.5.1) $\mathrm{ord}_S(x) = \max\{r\colon x \in m_S^r\}$

is a Euclidean algorithm on R_S, and for any constant $c > 0$ the function $(x \mapsto c^{-\mathrm{ord}_S(x)})$ from F to R is a valuation of F representing S. This should be compared with 4.1.6 since R_S is a Dedekind domain with unique maximal ideal; in fact it is a principal ideal domain. The completion $\hat{R}_S$ of R_S with respect to the m_S-adic topology, namely $\varprojlim_n R_S/m_S^n$, is the precisely ring of integers in the complete field F_S.

<u>Lemma 4.5.2</u> *Let S be a discrete spot on the field F such that F is complete and the residue field $F(S)$ is perfect, and has characteristic $p \neq 0$. There exists a unique function $j\colon F(S) \to R_S$ such that*

(4.5.2.1) $\rho j = 1$ *(where $\rho\colon R_S \to F(S)$ is reduction* mod S.

(4.5.2.2) $j(xy) = j(x)j(y)$ *for all $x,y \in F(S)$.*

If char(F) = p *, then j is a field homomorphism.*

If $\tilde{U}_S = \{x \in U_S\colon x^m = 1$ for some m with $(m,p) = 1\}$, then $\rho\colon \tilde{U}_S \to \mu(F(S))$ is an isomorphism with inverse j.

<u>Proof</u>   For $x \in F$ and n a positive integer let $U_n(X) = (\rho^{-1}(x^{p^{-n}}))^{p^n} \subset F$. The family $\{U_n(x),\, n \in N\}$ is a Cauchy filter[46] in F.   Define $j(x) = \lim_n U_n(x)$, then clearly j satisfies (4.5.2.1) and (4.5.2.2).   If $j'\colon F(S) \to R_S$ is any other function satisfying these conditions then $j'(x) \in U_n(x)$ for all n, so $j' = j$.   Clearly j is additive if char(F) = char(F(S)).

To prove the last part, it suffices to show that $\rho\colon \tilde{U}_S \to \mu(F(S))$ is injective.   Suppose then for some $x \in \tilde{U}_S$ of order $m \neq 1$ we have $\rho(x) = 1$. For any positive integer n there are integers a_n, b_n such that $a_n m + b_n p^n = 1$, since $(m,p) = 1$, and so x is the p^n-th power of x^{b_n}.

35

This means that x and 1 both belong to $U_n(1)$ for all n , hence $x = 1$.

In particular, under the conditions of 4.5.2 if $\operatorname{char}(F) = \operatorname{char}(F(S))$ and $F(S)$ is finite, then the map $j: F(S) \to F$ is an isomorphism onto the algebraic closure of the prime field in F .

<u>Definition 4.5.3</u>   If F is any field and S is a spot on F which is discrete, then a function $j: F(S) \to R_S$ satisfying (4.5.2.1) and (4.5.2.2) and the set $j(F(S))$ are both called a <u>system of multiplicative represent-</u><u>atives</u> (<u>SMR</u>) for $F(S)$.   If the SMR is unique, then we write $\operatorname{SMR}(S)$ for $j(F(S))$.

Suppose now that S is a discrete spot on F , and that F is complete with finite residue field.   Let $\eta \in m_S \backslash m_S^2$ be a prime element.   If $x \in m_S^n$ then it is easy to see that there is a unique element $a_n \in \operatorname{SMR}(S)$ such that $x - a_n \eta^n \in m_S^{n+1}$, namely $a_n = j\rho(x\eta^{-n})$, and hence by induction there is for each integer N a unique finite Laurent series $x_N = \sum\limits_{\operatorname{ord}_S(x)}^{N} a_i \eta^i$ with $a_i \in \operatorname{SMR}(S)$ such that $x - x_N \in m_S^{N+1}$ or equivalently $\operatorname{ord}_S(x - x_N) > N$. The sequence $\{x_N\}$ is Cauchy, and converges to x , so each element of F has a unique 'Laurent series' representation

$$(4.5.4) \qquad x = \sum\limits_{\operatorname{ord}_S(x)}^{\infty} a_i \eta^i \qquad \text{where } a_i \in \operatorname{SMR}(S) .$$

Such a Laurent series is integral (i.e. belongs to R_S) if and only if it is a power series.   This result shows that if $\operatorname{char}(F) = \operatorname{char}(F(S))$ then F is isomorphic to the field of Laurent series over $F(S)$, while R_S is isomorphic to the ring of power series over $F(S)$.

This also shows that if $F(S) = R_S/m_S$ is finite, then so is R_S/m_S^n for any $n \geq 1$, and so $\hat{R}_S$ is compact since $\hat{R}_S = \varprojlim\limits_{n} R_S/m_S^n$ is an inverse limit of finite, hence compact, spaces.

These methods admit straightforward generalisations to finite R_S-

algebras. Suppose that S is a discrete spot on the field F, that F is complete and that F(S) is perfect of characteristic $p \neq 0$. If A is a finite R_S-algebra, then[47] A is complete with respect to the m_S-adic topology and there exists an SMR($j: A/m_S A \to A$) as in 4.5.4. This enables us to prove <u>Hensel's Lemma</u>[48].

<u>Lemma 4.5.5</u> *Let S be a discrete spot on the field F such that F is complete, F(S) is perfect. Let f(X) be a monic polynomial in $R_S[X]$ such that $\rho(f) = g_1 g_2$ where g_1, g_2 are monic polynomials over F(S) and are relatively prime and $\deg \rho(f) = \deg f$. Then there exist monic polynomials $h_1, h_2 \in R_S[X]$ such that*

a) $f = h_1 h_2$

b) $\rho(h_i) = g_i$ *for* $i = 1, 2$.

<u>Proof</u>   Consider the finite R_S-algebra $A = R_S[X]/\langle \rho f \rangle$ and its residue algebra $A/m_S A = F(S)[X]/\langle \rho f \rangle$. Since $(g_1, g_2) = 1$ there is a decomposition $A/m_S A \cong F(S)[X]/\langle g_1 \rangle \oplus F(S)[X]/\langle g_2 \rangle$; let e_1 and e_2 be the idempotents in $A/m_S A$ corresponding under this decomposition to (1,0) and (0,1), and let $a_i = j(e_i) \in A$, where $j: A/m_S A \to A$ is an SMR. Since j is multiplicative, each a_i is idempotent and $a_1 a_2 = 0$, so there is a decomposition $A = Aa_1 \oplus Aa_2 \oplus A(1 - a_1 - a_2)$. The subalgebra $A(1 - a_1 - a_2)$ reduced mod M_S is zero, so is itself zero by Nakayama's lemma[49]. If $a \in A$ denotes the image of X under the map $R_S[X] \to A$ and $h_i(X) \in R_S[X]$ is the characteristic polynomial of $a: Aa_i \to Aa_i$, then $h_i(X)$ is monic and $h_1(X)h_2(X)$ is the characteristic polynomial of $a: A \to A$, which is f(X). Moreover $\rho h_i(X)$ is the characteristic polynomial of $\rho(a): F(S)e_i \to F(S)e_i$ which is $g_i(X)$.

4.6   RAMIFICATION

Let E/F be a finite extension and T a discrete spot on E extending

S on F. If v is a valuation on E representing T then $v(F^*)$ is a subgroup of $v(E^*)$ which has finite index since both groups are free of rank 1. The integer

$$(4.6.1) \qquad\qquad e_{T|S} = e = |v(E^*)/v(F^*)|$$

is independent of the choice of v and is called the <u>ramification index</u> of T over S, and satisfies (or is equally defined by) the conditions

$$(4.6.2) \qquad\qquad e.\mathrm{ord}_S(x) = \mathrm{ord}_T(x) \quad \text{for all} \quad x \in F;$$

(4.6.3)   if η is a prime of F at S and ξ of E at T then $\eta^{-1}\xi^e \in U_T$.

<u>Definition 4.6.4</u>   If S is a complete discrete spot on the field F and E/F is a finite extension with spot T extending S, then S is <u>unramified over</u> T if the ramification index $e_{T|S}$ is 1, and the extension $E(T)/F(S)$ is separable.

If S is a discrete spot on the field F which is not necessarily complete and E/F is a finite extension, then S is <u>unramified over</u> E if for each T extending S on F, the spot S on F_S is unramified over E_T.

From 4.3.7 we see that the ramification index is invariant under completion. The integer

$$(4.6.5) \qquad\qquad f_{T|S} = f = \deg(E(T)/F(S))$$

is called the <u>relative degree</u> of T over S.

<u>Proposition 4.6.6</u>   *If* E/F *is a finite extension,* T,S *are discrete spots on* E,F *which are also complete and* T|S *then*

$$e_{T|S}.f_{T|S} = \deg(E/F).$$

<u>Proof</u>   Let $a_1,\ldots,a_f$ be a basis for $E(T)/F(S)$ and let $j: E(T) \to E$ be an SMR for T, which exists by 4.5.2. If ξ is a prime element of E at T then we shall show that the set $D = \{j(a_r)\xi^s : 1 \le r \le f, 0 \le s < e\}$ is a

basis for E over F.

Let V be the F-subspace of E generated by D, which is complete, being a finite dimensional vector space over a complete field. If $x \in E$ and $\mathrm{ord}_T(x) = m$ then $x = u\xi^m = u'\eta^n\xi^s$ where $u,u' \in U_T$, and η is a prime element of F at S and $m = ne + s$ with $0 \leq s < e$. Let $x_1,\ldots,x_f$ be the elements of F(S) such that $\rho(u') = \Sigma x_i a_i$ and let $y = \Sigma j(x_i)j(a_i) \in V$. Now $x - y\eta^n\xi^s \in m_T^{m+1}$ since $u' - y \in m_T$, so there exists an element $v_1 \in V$ such that $\mathrm{ord}_T(x - v_1) > \mathrm{ord}_T(x)$. So by induction there is a Cauchy sequence $\{v_i\}$ in V which converges to x, so $x \in V$ and B generates E over F.

Suppose there is a non-trivial linear relation between the elements of D over F. We can assume without loss of generality that it is of the form

(4.6.6.1) $\Sigma d_{rs}j(a_r)\xi^s = 0$ where $d_{r,s} \in R_S$ and for some r,s we have $\mathrm{ord}_S(d_{r,s}) = 0$.

Let n be the least integer such that $\mathrm{ord}_S(d_{m,n}) = 0$ for some m. For any $s \neq n$, if $d_{r,s} \neq 0$ then we have

(4.6.6.2) $\mathrm{ord}_T(d_{r,s}j(a_r)\xi^s) = s + \mathrm{ord}_S(d_{r,s}).e > n$.

Multiplying (4.6.6.1) by ξ^{-n} and applying ρ we obtain $\Sigma_r \rho(d_{r,n})a_r = 0$, so $\mathrm{ord}_S(d_{r,n}) > 0$ for all r, which is a contradiction, so the elements of D are linearly independent over F.

<u>Corollary 4.6.7</u>   *If* E/F *is a finite separable extension and* S *is a discrete spot on* F *then*

$$\sum_{T|S} e_{T|S} f_{T|S} = \deg(E/F).$$

<u>Proof</u>   Immediate from 4.6.6 and 4.2.10.

<u>Remark 4.6.8</u>   Under the conditions of 4.6.6, the subspace K of E gener-

ated by SMR(T) is clearly a subfield of E whose residue field is also E(T).   The ramification index of K over F is 1.   This is a basic ingredient and starting point for class-field theory[50], along with the relation between the Galois groups of K/F and K(T)/F(S).

Suppose K/F is a finite normal extension of global fields with Galois group G, and that S is a discrete spot on F which is unramified over K. The group G acts transitively on the set S^K of spots that extend S by 4.2.4.   For $T \in F(K)$ extending S let H be the subgroup of G that stabilises T.   (If G is abelian, then H is independent of the choice of T, so depends only on S.)

Any element of H preserves the sets R_T, m_T and U_T, and so induces an automorphism of $K(T) = R_T/m_T$ over F(S).   This defines a homomorphism $H \to G_T = \mathrm{Gal}(K(T)/F(S))$.

If $e \in K$ is a primitive element which belongs to U_T, then its minimum polynomials f over F and g over F_S both have coefficients in R_S, and f = gh for some $h \in F_S[X]$.   The degree of g is equal to d, the relative degree of T over S, since S is unramified.   The reduction $\bar{e}$ of e mod S generates K(T) over F(S), so its minimum polynomial over F(S) also has degree d.   Since $(\rho g)(\bar{e}) = \rho(g(e)) = 0$, we see that ρg is the minimum polynomial of $\bar{e}$.   By Hensel's lemma (4.5.5) the map ρ induces a bijection between the roots of g in K_T and the roots of ρg in K(T).

Now suppose $\sigma \in G_T$.   The element $\sigma\bar{e}$ is also a root of ρg, so there is a unique root e' of g such that $\sigma(\bar{e}) = \rho(e')$.   Since e' is a root of f, there is a unique $\sigma' \in G$ such that $e' = \sigma(e)$.   Since σ' permutes the roots of g, it preserves the spot T and so belongs to H.   Thus we have shown:

<u>Proposition 4.6.9</u> *If K/F is a finite normal extension of global fields with Galois group G, and S is a discrete spot on S which is unramified over F then the reduction homomorphism*

$$\{\sigma \in G:\ \sigma(T) = T\} \to \mathrm{Gal}(K(T)/F(S))$$

is an isomorphism for each T extending S.

As an illustration of these somewhat technical results, we shall describe in some detail their interpretation for the p-adic numbers.

<u>Example 4.6.10</u> By definition, the field Q_p of p-adic numbers is the completion of Q at the p-adic spot, and the ring Z_p of p-adic integers is its corresponding valuation ring. By 4.3.7 the residue field of Q_p is the field F_p with p elements. The results of 4.5.5 confirm the standard presentation of Q_p as certain 'Laurent series' in p.

By 4.5.2 we see that Z_p contains (p-1)-th roots of unity since F_p does. By 4.3.7 the residue field of Q_p^a is F_p^a. The algebraic closure of Q in Q_p^a is the algebraic closure of Q which can also be considered[51] as the field of algebraic numbers in C. Let K_n be the subfield of Q^a generated by the (p^n-1)-th roots of unity, and let $K_n' \subset Q_p^a$ be its completion. Let L_n be the residue field of K_n' (and of K_n).

If ω is a primitive (p^n-1)-th root of unity in Q^a, then ω generates K_n over Q and K_n' over Q_p. Since the powers of ω are roots of unity of order prime to p, their images in L_n are distinct by 4.5.2. In particular, this means that $\bar{\omega} = \rho(\omega)$ generates a subfield of order p^n (since $\omega^{p^m-1} - 1 \neq 0$ for $0 < m < n$).

If m(X) is the minimum polynomial of ω over Q_p, then m(X) belongs to $Z_p[X]$ since ω is integral over Z_p (see 4.4.1). The roots of m(X) are distinct powers of ω, so the roots of $\rho m(X) \in F_p[X]$ are distinct powers of $\bar{\omega}$, so $\rho m(X)$ is irreducible (by Hensel's Lemma). Thus $\rho m(X)$

41

is the minimum polynomial of $\bar{\omega}$ over F_p, which has degree n. Thus $\deg(K_n'/Q_p) = m(X) = \deg \rho\, m(X) = n$, so by 4.6.6 we have $L_n = F_p(\bar{\omega})$ is a field of order p^n, and K_n'/Q_p is unramified $(e = 1)$.

In addition this shows that $m(X) = \prod_{i=0}^{n-1}(X - \omega^{p^i})$ and that the Galois group of K_n'/Q_p is cyclic of order n generated by the Frobenius map $\pi: K_n' \to K_n'$ which takes ω to ω^p.

(Moreover this shows that for any finite field F there is an algebraic number field K and a spot S on K whose residue field is F, and an automorphism of K preserving S which induces the Frobenius map on F. This fact is used to lift problems over finite fields to the complex numbers[52].)

If x is an element of F_p^a, then $x \in L_n$ for some n, so we can define $j(x) = j_n(x) \in Q_p^a$, where $j_n: L_n \to K_n'$ is the SMR for the field K_n'. This is independent of the choice of n since if n divides m (so that L_n is contained in L_m) then j_m restricted to L_n is j_n by uniqueness (see 4.5.2). The function $j: F_p^a \to Q_p^a$ is an SMR for the field Q_p^a. The elements of $j(F_p^a{*})$ are the m-th roots of unity where m is relatively prime to p, and so considered as complex numbers lie on S^1. Thus the map j defines at the same time a monomorphism from $F_p^a{*}$ into $Q_p^a{*}$ and a monomorphism from $F_p^a{*}$ into S^1.

The field Q_p^a is not complete. Once can show[53] that for any $\alpha \in Q_p^a$ there is a neighbourhood U of α such that $\beta \in U$ implies $Q_p(\alpha) \subset Q_p(\beta)$. By using this fact it is easy to see that if a_n is a p_n-th root of p then the series $\Sigma a_n p^n$ cannot converge in Q_p^a, even though it is Cauchy. Let Ω_p denote the completion of Q_p^a, which is algebraically closed, and has residue field F_p^a by 4.3.7. The field Ω_p for p-adic analysis plays the analogous role to C for classical analysis. The SMR for Ω_p provides a link between corresponding unit discs in the form of the map

$$\{z \in \Omega_p : |z| \le 1\} \stackrel{\rho}{\to} F_p^a \stackrel{j}{\to} \{z \in C : |z| \le 1\}.$$

4.7 NORMALISED VALUATIONS

We now consider the theory of valuations applied to the global fields. Let K be a global field, $P(K)$ the set of non-trivial spots on K, and $S \in P(K)$.

If S is discrete, let q_S denote the order of the finite residue field $K(S)$. The valuation

$$(4.7.1) \qquad\qquad x \mapsto q_S^{-\mathrm{ord}_S(x)}$$

on K which represents the spot S (see 4.5) is called the <u>normalised valuation</u> of S, and is denoted by $x \mapsto |x|_S$.

If on the other hand S is archimedean (so that K is a number field) then S extends the infinite spot on Q, and K_S is a finite extension of R. There are two cases to consider.

(4.7.2.1) <u>Case 1</u> The inclusion $R \to K_S$ is a continuous algebraic isomorphism. In this case S is called a <u>real</u> spot, and the valuation on K_S (and by restriction on K) which corresponds under this isomorphism to the absolute value $x \mapsto |x|$ on R is denoted by $x \mapsto |x|_S$ and is called the <u>normalised valuation</u> of S and does represent the spot S.

(4.7.2.2) <u>Case 2</u> The inclusion $R \to K_S$ is continuous, and is algebraically isomorphic to the inclusion $R \to C$. By uniqueness of valuations on finite extensions of complete fields (4.2.1) it follows that there is a continuous algebraic isomorphism from K_S to C which is the identity on R. The function $x \mapsto |x|_S$ from K_S to R which corresponds under this isomorphism to the function $x \mapsto |x|^2$ on C is not a valuation as it fails to satisfy (4.1.1.2). However it does satisfy the inequality

$$|x + y|_S \le 2(|x|_S + |y|_S) \quad \text{for all} \quad x,y \in C$$

or equivalently the function $x \mapsto |x|_S^r$ is a valuation on K_S for $0 < r \leq 1/2$ and any such valuation represents the spot S. In this case, S is called a <u>complex</u> spot, and the function $x \mapsto |x|_S$ is often called the <u>normalised valuation</u> of S.

<u>Lemma 4.7.3</u> *Let K be a global field and L/K   a finite separable extension. If $S \in P(K)$ and $T|S$ then*

$$|x|_T = |N_{L_T/K_S}(x)|_S \quad \text{for all} \quad x \in L_T.$$

<u>Proof</u> If S is archimedean then L_T/K_S is isomorphic to one of the extensions $R/R, C/R$ or C/C in which case the result is trivially true.

Suppose then that S is discrete. By (4.2.1) we know that there is a real number $r > 0$ such that $|x|_T = |N_{L_T/K_S}(x)|_S^r$ for all $x \in L_T$. Let f be the relative degree $\deg(L(T)/K(S))$ and e the ramification index of T over S, so that $q_T = q_S^f$ and $\mathrm{ord}_T(x) = e.\mathrm{ord}_S(x)$.

If x is an element of K_S then we have

$$|x|_T = q_T^{-\mathrm{ord}_T(x)} = q_S^{-ef.\mathrm{ord}_S(x)} = |x|_S^n \quad \text{since} \quad ef = n \quad \text{by 4.6.6.}$$

But in this case, $N_{L_T/K_S}(x) = x^n$, so it follows $r = 1$.

<u>Corollary 4.7.4</u> *If K is a global field, L/K a finite separable extension and $S \in P(K)$, then*

$$\prod_{T|S} |x|_T = |N_{L/K}(x)|_S.$$

<u>Proof</u> Immediate from 4.7.3 and 4.2.11.

This leads to the <u>Artin-Whaples formula</u>[54].

<u>Theorem 4.7.5</u> *If K is a global field, then for any $x \in K^*$ we have*

(4.7.5,1) $|x|_T = 1$ *for almost all* $T \in P(K)$

(4.7.5.2) $\prod_{T \in P(K)} |x|_T = 1$

<u>Proof</u> By definition, K is a finite separable extension of F, where F is

44

either Q of $F_p(X)$. As usual the proof follows from the result for $K = F$.

If x is an element of K, then there exist elements $a_0,\ldots,a_{n-1}$ such that $x^n + a_{n-1}x^{n-1} + \ldots + a_0 = 0$. Since (4.7.5.1) is trivially true for $K = F$ (by 4.1.8 or 4.1.9) we see that $\{a_0,\ldots,a_{n-1}\} \subset R_S$ for almost all spots $S \in P(F)$. From 4.4.1 and the fact that each element of $P(F)$ admits only a finite number of extensions to K by (4.2.3) we see that $x \in R_T$ for almost all $T \in P(K)$. Similarly $x^{-1} \in R_T$ for almost all $T \in P(K)$, which proves (4.7.5.2).

By using 4.7.3 we see that

$$\prod_{T \in P(K)} |x|_T = \prod_{S \in P(F)} \prod_{T|S} |x|_T$$

$$= \prod_{S \in P(F)} |N_{K/F}(x)|_S$$

$$= 1 \quad \text{since (4.7.5.2) is trivially true}$$

for $K = F$.

Thus there is an 'infinite relation' between the spots on K. However, there can be no finite relation between them, as we now show by using the Weak Approximation Theorem[55]:

<u>Theorem 4.7.6</u> *Let* $v_1,\ldots,v_n$ *be a finite set of inequivalent non-trivial valuations on a field* F. *Then for any* $x_1,\ldots,x_n \in F$ *and any* $\varepsilon > 0$ *there exists an* $x \in F$ *such that* $v_i(x-x_i) < \varepsilon$ *for all* i.

It is easy to show that such an x can be chosen non-zero.

<u>Corollary 4.7.7</u> *Let* $v_1,\ldots,v_n$ *be a finite set of inequivalent non-trivial valuations on a field* F. *If* $c_1,\ldots,c_n$ *are real numbers such that*

$$\prod_{i=1}^{n} v_i(x)^{c_i} = 1 \quad \text{for all } x \in F*$$

then $c_i = 0$ *for all* i.

<u>Proof</u> Let $a = \sum_{c_i > 0} c_i$ and $b = -\sum_{c_i > 0} c_i$; assume without loss of

generality that a is non-zero. Let ε be a real number such that $0 < \varepsilon < 1$ and $\varepsilon^a(1-\varepsilon)^{-b} < 1$. By 4.7.6 there exists an $x \in F^*$ such that

$$(4.7.7.1) \qquad v_i(x) < \varepsilon \qquad c_i > 0,$$

$$\text{and } v_i(x-1) < \varepsilon \text{ if } c_i < 0.$$

If $c_i < 0$ then we have $v_i(1) - v_i(x) < \varepsilon$, so $v_i(x) > 1 - \varepsilon$, i.e.

$$(4.7.7.2) \qquad v_i(x)^{c_i} < (1-\varepsilon)^{c_i} \quad \text{for } c_i < 0.$$

Putting together (4.7.12.1) and (4.7.12.2) we obtain

$$\prod_i v_i(x)^{c_i} < \varepsilon^a(1-\varepsilon)^{-b} < 1 \quad \text{which is a contradiction.}$$

The Artin-Whaples formula enables us to prove the following important result.

<u>Proposition 4.7.8</u> *If* L/K *is a finite separable extension of global fields, then almost all spots on* K *are unramified.*

<u>Proof</u> By taking the normal closure of L/K if necessary, we can assume that L/K is a normal extension, since a spot on K which is unramified on L is a fortiori unramified over any subfield of L over K. Since all residue fields are finite, all extensions of residue fields are separable.

Let $\alpha_1 \in L$ be a primitive element with minimum polynomial $f(X)$ of degree $n = \deg(L/K)$. Let $\alpha_1, \ldots, \alpha_n$ be the distinct roots of $f(X)$ in L. By 4.7.5 $f(X) \in R_S[X]$ for almost all spots $S \in P(K)$ and so $\{\alpha_1, \ldots, \alpha_n\} \subset R_T$ for almost all $T \in P(L)$. Suppose that some such S is ramified so that $r = \deg(L(T)/K(S))$ is less than $m = \deg(L_T/K_S)$. Let $f_1(X)$ be the minimum polynomial of α_1 over K_S so that $\deg f_1(X) = m$ and the roots of $f_1(X)$ are in the set $\{\alpha_1, \ldots, \alpha_n\}$. Let $\bar{g}(X)$ be the minimum polynomial of $\rho(\alpha_1)$ over $K(S)$ which has degree less than or equal to r, so that $\rho f(X) = \bar{g}(X)\bar{h}(X)$ where $h(X)$ is some non-constant polynomial over $K(S)$.

By Hensel's Lemma (4.5.5), $g(X)$ and $h(X)$ have a common factor since f_1 is irreducible over K_S, and since the roots of $\bar{g}(X)$ and $\bar{h}(X)$ (in $L(T)$) are of the form $\rho(\alpha_i)$ for some i, it follows that $\rho(\alpha_i) = \rho(\alpha_j)$ for some i,j with $i \neq j$. If $D = \prod_{i \neq j} (\alpha_i - \alpha_j) \in R_T$ then $\rho(D) = 0$, and so $D \in m_T$. The element D is non-zero since the roots of $f(X)$ are distinct.

Hence we have shown that if a spot T on L is ramified, then either $\{\alpha_0, \ldots, \alpha_{n-1}\} \not\subseteq R_S$ (where S is T restricted to K) or $\{\alpha_0, \ldots, \alpha_{n-1}\} \subseteq R_S$ and $D \in m_T$. In either case the set of possibilities is finite by 4.7.5 which proves the proposition.

<u>Remark 4.7.9</u> The element D defined in 4.7.8 is invariant under the Galois group of L/K so lies in K. It is called the <u>discriminant</u> of the polynomial $f(X)$. The previous proposition thus shows that if $f(X) \in R_S[X]$ and S is ramified, then $|D|_S < 1$.

<u>Example 4.7.10</u> Consider a quadratic extension $Q(\sqrt{d})/Q$ where d is a non-zero square-free integer. The minimum polynomial $X^2 - d$ of $\sqrt{d}$ has discriminant $4d$ and so the only possible ramified spots on Q are those p-adic spots for which p divides d, or $p = 2$.

If $p | d$, then $|\sqrt{d}|_p^2 = |d|_p = 1/p$, so $|\sqrt{d}|_p = 1/\sqrt{p}$ so this spot is ramified $(e = 2)$.

Suppose now that d is odd; consider the extension $Q_2(\sqrt{d})/Q_2$. Since[56] the rational number $\binom{1/2}{r} 4^r$ is an integer for $r \in N$, the Q_2 power series $\sum \binom{1/2}{r} 8^r x^r$ converges for $x \in Z_2$ to a square root of $1 + 8x$. Write $d = a + 8k$ where $1 \leq a \leq 7$ and a is odd, so that $d = a(1 + 8x)$ where $x = k/a \in Q_2$. We see that $\sqrt{d} \in Q_2$ if and only if $\sqrt{a} \in Q_2$. A simple calculation in 2-adic power series shows that $\sqrt{3}, \sqrt{5}, \sqrt{7}$ do not belong to Q_2, that is to say[57] for d odd, $\sqrt{d} \in Q_2$ if and only if $d \equiv 1 \bmod 8$.

The integral closure of Z_2 in $Q_2(\sqrt{d})$ (or indeed of Z_p in $Q_p(\sqrt{d})$)

can easily be determined exactly as one determines the integral closure of Z

in $Q(\sqrt{d})$.   If $d \equiv 3 \bmod 4$,   the ring of integers in $Q_2(\sqrt{d})$ is $Z_2(\sqrt{d})$

and the residue field is thus F_2,   so the extension is ramified $(e = 2)$.

If $d \equiv 5 \bmod 8$,   the ring of integers on $Q_2(\sqrt{d})$ is $Z_2(w)$ where

$w = 1/2(1 + \sqrt{d})$ has minimum polynomial $X^2 + X + (1 - d)/4$.   Since

$(1 - d)/4 \equiv 1 \bmod 2$,   the image of w in the residue field satisfies the

equation $X^2 + X + 1 = 0$,   so the residue field is F_4,   and the extension

$Q_2(\sqrt{d})/Q_2$ is unramified.

<u>Example 4.7.11</u>   Consider the extension $K/F_p(X)$ where K is the splitting

field of the polynomial $Y^2 - f$ over $F_p(X)$,   and $p \nmid 2$.   The discriminant

of this polynomial is $4f(X)$,   and as in the previous example a P-adic spot

(where P is a maximal ideal of $F_p[X]$) is ramified if and only if $f(X) \in P$.

The infinite spot is unramified or ramified according as $\deg(f)$ is even or

odd.

   Let K be an algebraic function field of characteristic $p \neq 0$.   If

$S \in P(K)$,   then the order q_S of the residue field $K(S)$ is a power of p.

The <u>degree</u> of S, $\deg(S)$,   is defined to be $\log_p(q_S)$.

   If $K/F_p(X)$ is a finite separable extension of degree n,   then the

rational number

$$(4.7.12) \qquad g = 1/2\left(\sum_{S \in P(F_p(X))} \deg(S) \sum_{T \in S}^{K} (e_{T|S} - 1) \right) + 1 - n$$

depends only on K and not the extension $K/F_p(X)$,   and is called the <u>genus</u>

of K.   It is in fact a non-negative integer[58].   A simple calculation from

example 4.7.11 shows that this definition agrees with that given in §3.3.

4.8   APPENDIX: LOCALLY COMPACT FIELDS

Let F be a locally compact field, and ν a Haar measure on the additive group of F.   If x is a non-zero element of F then the function

$X \mapsto \nu(xX)$ is also a Haar measure on the measureable sets of F; so there

is a real number $\mathrm{mod}_F(x)$ such that

$$\nu(xX) = \mathrm{mod}_F(x)\,\nu(X) \quad \text{for all measureable sets X.}$$

If we define $\mathrm{mod}_F(0) = 0,$ then the function $\mathrm{mod}_F: F \to R$ is independent of

the choice of ν and is called the <u>modular function</u> of F.   It has the

following properties[59]:

(4.8.1) $\mathrm{mod}_F(x) \geq 0$ for all $x \in F$, and $\mathrm{mod}_F(x) = 0$ if and only if

        $x = 0;$

(4.8.2)   there exists an $A > 0$ such that $\mathrm{mod}_F(x+y) \leq A(\mathrm{mod}_F(x) + \mathrm{mod}_F(y))$

        for $x,y \in F$;

(4.8.3) $\mathrm{mod}_F(xy) = \mathrm{mod}_F(x)\,\mathrm{mod}_F(y)$ for $x,y \in F$;

(4.8.4) $\mathrm{mod}_F: F \to R$ is continuous;

(4.8.5)   if E/F is a finite separable extension of locally compact fields,

then

$$\mathrm{mod}_E(x) = \mathrm{mod}_F(N_{E/F}(x)) \quad \text{for all } x \in E.$$

Thus there is a constant $c > 0$ such that $x \mapsto \mathrm{mod}_F(x)^c$ is a valua-

tion on F.   The following is the classification theorem for locally compact

fields[60].

<u>Theorem 4.8.6</u>   *If F is a non-discrete locally compact field with modular*

function $\mathrm{mod}_F: F \to R$, then F is a finite separable extension of the locally

compact field

(4.8.6.1) R <u>if</u> $\mathrm{mod}_F(m) \geq 1$ <u>for all</u> $m \in N$; and char F=0

(4.8.6.2) Q_p <u>if</u> $0 < \mathrm{mod}_F(p) < 1$; and char F= 0

(4.8.6.3) $F_p((X))$ <u>if</u> char $(F) = p \neq 0$.

In each case the topology on F is that induced by the spot S represented by the valuation $x \mapsto \mathrm{mod}_F(x)^c$ *for some suitable constant* $c > 0$. *If this is non-archimedean, then then valuation ring* R_S *is the maximal compact subring of* F.

<u>Corollary 4.8.7</u> *If* K *is a global field and* S *is a non-trivial spot on* K *then the modular function* $\mathrm{mod}_{K_S} : K_S \to R$ *is the normalised valuation, i.e.* $\mathrm{mod}_{K_S}(x) = |x|_S$ *for all* $x \in K_S$.

<u>Proof</u> By 4.7.3 and 4.8.5 it suffices to prove the result for $K = R$, Q_p or $F_p((X))$. It is trivially true for $K = R$. We shall prove the result for $K = Q_p$, the case when $K = F_p((X))$ being similar.

We know by (4.8.6) that there is a constant $c > 0$ such that $\mathrm{mod}_{Q_p}(x) = |x|_p^c$ for all $x \in Q_p$. When $x = p$ we have $|p|_p = 1/p$. Now Z_p is a compact subspace of Q_p, hence measureable, and since pZ_p has index p in Z_p, we have $\mathrm{mod}_{Q_p}(p) = 1/p$, so $c = 1$ as required.

5 Global fields

In this chapter we describe the algebraic theory of the ring of integers in a global field from the point of view of valuations, thus dealing with algebraic number fields and algebraic function fields simultaneously.   In order to do this we introduce some notation which will be fixed throughout this chapter.

Let F denote either Q or $F_p(X)$, and let K/F be a finite separable extension of degree n, so that K is a typical global field. Let

$A \subset F$ be the Euclidean domain $\begin{cases} Z & \text{if } F = Q \\ F_p[X] & \text{if } F = F_p(X) \ ; \end{cases}$

B be the integral closure of A in K, k the algebraic (= integral) closure of F_p in K if char $K = p \neq 0$, and q its order ;

U be the group of units in B ;

∞ be the infinite spot on F ;

$P = P(K)$ be the set of non-trivial spots on K ;

$P_\infty = P_\infty(K/F) = \{S \in P : S \mid \infty\} = \{\infty_1,\ldots,\infty_r\}$;

$P_0 = P \setminus P_\infty$;

$P_a = \{S \in P : S$ is archimidean$\}$;

$P_{na} = \{S \in P : S$ is non-archimidean$\}$.

We emphasize that in the case char $K \neq 0$ the ring B is determined by the embedding $F \to K$, not just by K, as has been indicated by example 3.5.6.   The ring homomorphism $A \to B$ gives B the structure of an A-module.

<u>**Proposition 5.1.1**</u>   B *is a free* **A**-*module of rank* $n = \deg (K/F)$.

<u>**Proof**</u>   Let $\{b_1,\ldots,b_n\} \subset B$ be a basis for K as an F vector space, and let $\{d_1,\ldots,d_n\}$ be a dual basis with respect to the non-singular bilinear form $(x,y) \rightarrow Tr_{K/F}(xy)$ on K, so that $Tr_{K/F}(b_i d_j) = \delta_{i,j}$.   If b is any element of B then $b = \Sigma x_j d_j$ where $x_j \in F$.   Now for any index i, the element bb_i belongs to B and so $Tr_{K/F}(bb_i)$ belongs to A (since the trace of an element is a coefficient, up to sign, in its minimal polynomial) ;   but clearly

$$Tr_{K/F}(bb_i) = Tr_{K/F}(\sum_j x_j d_j b_i) = \sum_j x_j Tr_{K/F}(d_j b_i) = x_i .$$

Hence B is contained in the sub-A-module of K generated by $\{d_1,\ldots,d_n\}$ and so is finitely generated.   Since B is torsion-free as an A-module it is free and has rank equal to $\dim_F(K \otimes_A B) = \deg(K/F)$.

5.2   IDEAL THEORY

<u>**Proposition 5.2.1**</u>   *The relation between the ring* B *and the spots on* K *is as follows:*

(5.2.1.1) $\qquad B = \bigcap_{S \in P_0} R_S$;

(5.2.1.2) $\qquad if$ char $K \neq 0$ *then* $k = \bigcap_{S \in P} R_S$.

<u>**Proof**</u>   Both parts of the proposition follow from 4.4.1 using the fact that 5.2.1 is true for K = F by either 4.1.8 or 4.1.9 .

For any spot $S \in P_0$ the ring B is contained in R_S so the ideal $p_S = m_S \cap B$ is a prime ideal of B.   We shall show presently (in 5.2.8) that p_S is a maximal ideal and moreover every maximal ideal is of this form.   First we need some more facts about valuations, starting with a trivial result.

<u>**Proposition 5.2.2**</u>   *Let* R *be a ring and* M_1, M_2 *distinct maximal ideals of*

R. *For any positive integer* $\mathbf{m}$ *there exists an element* $\mathbf{a} \in R$ *such that* $\mathbf{a} \in M_1^m$ *and* $\mathbf{a} - 1 \in M_2^m$.

Since every spot in P_0 is discrete, we have:

<u>Corollary 5.2.3</u> *If* $\mathbf{S,T}$ *are distinct spots in* P_0, *then for any* $\varepsilon > 0$ *there exists an element* $\mathbf{x} \in B$ *such that* $|\mathbf{x}|_S < \varepsilon$ *and* $|\mathbf{x} - 1|_T < \varepsilon$.

From this corollary one can prove the following result, known as the <u>Strong Approximation Theorem</u>[61].

<u>Theorem 5.2.4</u> *Let* $\mathbf{X}$ *be a finite subset of* P_0 *and suppose that for each* $\mathbf{S} \in \mathbf{X}$ *we are given an element* $\mathbf{x}_S \in K$. *For any* $\varepsilon > 0$ *there exists an element* $\mathbf{x} \in B$ *such that* $|\mathbf{x} - \mathbf{x}_S|_S < \varepsilon$ *for all* $\mathbf{S} \in \mathbf{X}$.

For any ideal $\mathbf{I}$ of B and spot $\mathbf{S} \in P_a$ define

$$\mathrm{ord}_S(I) = \max\{m \in Z : I \subset p_S^m\} \quad \text{and} \quad |I|_S = \max\{|\mathbf{x}|_S : \mathbf{x} \in I\}.$$

The following facts are easily verified[62].

(5.2.5.1) $\qquad |I|_S = q_S^{-\mathrm{ord}_S(I)}$;

(5.2.5.2) $\qquad |B\mathbf{x}|_S = |\mathbf{x}|_S$ for any $\mathbf{x} \in B$;

(5.2.5.3) $\qquad |I|_S \geq 0$, and $|I|_S = 0$ if and only if $I = 0$;

(5.2.5.4) $\qquad |I + J|_S \leq \max\{|I|_S , |J|_S\}$;

(5.2.5.5) $\qquad |IJ|_S = |I|_S , |J|_S$;

(5.2.5.6) $\qquad$ for any ideal $\mathbf{I}$ of B we have $|I|_S = 1$ for almost

$\qquad\qquad$ all $\mathbf{S} \in P_a$.

Using the strong approximation theorem one can show[63] :

<u>Proposition 5.2.6</u> *If* $\mathbf{I}$ *is an ideal of* B *then*

$$I = \{\mathbf{x} \in B : |\mathbf{x}|_S \leq |I|_S \text{ for all } \mathbf{S} \in P_0\}.$$

If $\mathbf{I,J}$ *are two ideals of* B *then* $I \subseteq J$ *if and only if* $|I|_S \leq |J|_S$ *for all* $\mathbf{S} \in P_0$.

$\underline{\text{Corollary 5.2.7}}$ *If* M *is a maximal ideal of* B, *then* $M = p_S$ *for some*

unique $S \in P_0$.

$\underline{\text{Proof}}$ It follows from 5.2.6 that if I is an ideal of B such that

$|I|_S = 1$ for all $S \in P_0$, then $I = B$.   Thus for some $S \in P_0$ we have

$|M|_S < 1$ so $M \subset p_S$ and hence $M = p_S$.   Uniqueness is trivial.

It is now a simple step to the unique factorisation theorem.

$\underline{\text{Theorem 5.2.8}}$ *If* I *is a proper ideal of* B *then*

$$I = \prod_{S \in P_0} p_S^{\,\text{ord}_S(I)}$$

is the unique factorisation of I *as a product of maximal ideals.*

$\underline{\text{Proof}}$ Since $|p_S|_T = 1$ if $S \neq T$, uniqueness is trivial.   To prove

existence, let $J = \prod_{S \in P_0} p_S^{\,\text{ord}_S(I)}$, a finite product by (5.2.5.6).   Clearly

we have $I \subset J$.   Now for any spot $T \in P_0$ we have

$$|J|_T = \prod_{S \in P_0} \left| p_S \right|_T^{\,\text{ord}_S(I)} = \left| p_T \right|_T^{\,\text{ord}_T(I)} = |I|_T$$

and so $J = I$ by 5.2.6.

This theorem has the following immediate consequences:

(5.2.9.1)   the ring B is a Dedekind domain, i.e. is integrally closed and
every non-zero prime ideal is maximal;

(5.2.9.2)   every spot $S \in P_0$ is an M-adic spot (see 4.1.6) for some unique
maximal ideal M of B such that $K(S) = B/M$ and moreover B/M
is finite;

(5.2.9.3)   the monoid of non-zero ideals of B under multiplication is the
free monoid on the maximal ideals;

(5.2.9.4)   every non-zero ideal of B is (as a sub-A-module of B) a free
A-module of rank n.

5.3   UNITS IN B

From 5.2.1 it follows that $U = \bigcap_{S \in P_0} U_S$.

54

<u>**Lemma 5.3.1**</u> *The group* $\underset{S \in P}{\cap} U_S$ *is the group* $\mu(K)$ *of roots of unity in* K.

<u>**Proof**</u>   If char(K) $\neq$ 0 then we have

$$k^* = \mu(K) \subset \underset{S \in P}{\cap} U_S \subset (\underset{S \in P}{\cap} R_S) \setminus \{0\} = k^* \text{ using (5.2.1.2) .}$$

If char(K) = 0 clearly we have $\mu(K) \subset \underset{S \in P}{\cap} U_S$.   Suppose that $x \in \underset{S \in P}{\cap} U_S$.   If P_∞ contains a real spot, say ∞_1 , then $|x|_{\infty_1} = 1$ implies that $x = \pm 1$ so that x belongs to $\mu(K)$.   If all the spots in P_∞ are complex, then the image of x , say x' , in the ring $\underset{i=1}{\overset{r}{\Pi}} K_{\infty_i} = R$ (which is isomorphic as a topological ring to C^r) lies on the r-dimensional torus $T^r = \{(x_1,\ldots,x_r) \in R : |x_i|_{\infty_i} = 1\}$ which is a compact group.   Hence the set of powers of x' is either finite (which implies that x is a root of unity) or has the element 1 as a limit point.   We show that this latter case cannot happen.

Assume for a contradiction that 1 is a limit point of powers of x' . Hence there exists an integer m such that $|x^m - 1|_\infty < 1/2$ for $i = 1,\ldots,r$.   But for any spot $S \in P_0$ we have $x^m - 1 \in R_S$ since $x \in U_S \subset R_S$ and S is non-archimidean.   If $z = x^m - 1$ then we have

$$1 = \underset{S \in P}{\Pi} |z|_S = \left(\underset{i=1}{\overset{r}{\Pi}} |z|_{\infty_i} \right) \cdot \left(\underset{S \in P_0}{\Pi} |z|_S \right) < 1/2^r ,$$

which is a contradiction.

This result is part of the <u>Dirichlet Unit Theorem</u>[64].

<u>**Theorem 5.3.2**</u>   *The group* U *of units in* B *is isomorphic to* $Z^{r-1} \oplus \mu(K)$ *where* $r = |P_\infty|$.

<u>**Sketch proof**</u>   Consider the map $g : U \to R^r , x \to \log|x|_{\infty_1},\ldots,\log|x|_{\infty_r}$. By the Artin-Whaples formula 4.7.5, the group Im(g) lies in the hyperplane $H = \{(X_1,\ldots,X_r) \in R^r : \underset{i=1}{\overset{r}{\Sigma}} X_i = 0\}$, and since it is a discrete subgroup is free of rank less than or equal to r - 1 .   By means of the strong approximation theorem 5.2.4 it can be shown [65] that Im(g) contains a basis

for H and so has rank equal to $r - 1$.   Since the sequence

$O \to \mu(K) \to U \overset{g}{\to} \text{Im}(g) \to 0$ is exact by 5.3.1 and splits, the result follows.

<u>Definition 5.3.3</u>   A set $\{e_1, \ldots, e_{r-1}\}$ of elements of U is called a

<u>fundamental set of units</u> if the set $\{g(e_1), \ldots, g(e_{r-1})\}$ is a basis for

Im(g) .   Let $\delta = (\delta_1, \ldots, \delta_r) \in R^r$ be the vector with $\delta_i = 2$ if ∞_i is a

complex spot, but $\infty_i = 1$ otherwise.   Put $d = \sum_{i=1}^{r} \delta_i$.   The real number

$(1/d)|\det \Delta|$,   where Δ is the $r \times r$ matrix whose first row is δ and whose

$(i + 1)$-th row for $i = 1, \ldots, r - 1$ is (e_i) ,   is independent of the choice

of fundamental set of units and is called the <u>regulator</u> of K/F,   denoted

by R .

We note that the number d is equal to $n = \deg(K/F)$ if char(K) = 0

but is equal to $r = |P_\infty|$ if char(K) $\neq$ 0 .   Moreover, if char(K) = p $\neq$ 0

then R is an integral multiple of $(\log q)^{r-1}$ since normalised valuations

are powers of q .

By adding each of the first $(r - 1)$ columns of A to the last column

and using the fact that $g(e_i) \in H$, we see that

$$R = |\det \tilde{\Delta}| \quad \text{where} \quad \tilde{\Delta} \quad \text{is the} \quad (r - 1) \times (r - 1) \quad \text{matrix}$$

$$\text{with} \quad \tilde{\Delta}_{ij} = \log |e_i|_{\infty_j} .$$

5.4   THE IDEAL CLASS GROUP

As we have already observed in (5.2.9.3) the set of non-zero ideals of B

form a free semigroup with respect to multiplication.   Let Frac(B) denote

the associated abelian group written additively, so that Frac(B) is $Fr(P_0)$,

the free abelian group on the set P_0 .   The elements of Frac(B) can also

be identified naturally with the fractional ideals[66] of B.

The function which associates to each non-zero element x of B the

principal ideal generated by x induces a homomorphism $\phi : K^* \to$ Frac(B)

whose kernel is clearly the group U .

56

Definition 5.4.1   The cokernel of $\phi : K^* \to \mathrm{Frac}(B)$ is called the <u>ideal class group</u> of B , and is denoted by $\mathrm{ICG}(B)$.   Its order is called the <u>class number</u> of B , and is written h or h_B .

Clearly the ring B is a principal ideal domain if and only if $h_B = 1$.

The ideal class group has various other important descriptions.   We mention one which we shall need later.   Let $\mathrm{Pic}(B)$ denote the group of (isomorphism classes of) <u>locally-free</u>[67] B-<u>modules of rank</u> 1 with respect to tensor product over B .

<u>Proposition 5.4.2</u>   *The groups* $\mathrm{ICG}(B)$ *and* $\mathrm{Pic}(B)$ *are naturally isomorphic.*

<u>Proof</u>   Suppose I is a non-zero ideal of B .   We show that I is locally-free of rank 1 as a B-module.   Let P be a prime ideal of B ;   the inclusion $B \hookrightarrow B_P$ induces a monomorphism

$$I = B \otimes_B I \longrightarrow B_P \otimes_B I = I_P$$

so that I_P is a non-zero ideal of R_P .   But B_P is a principal ideal domain (see 4.5.1) so I_P is a free R_P-module of rank 1 .

Moreover, since the natural map $I \otimes_B J \longrightarrow IJ$ (where I,J are ideals of B) is an isomorphism of B-modules, we have a homomorphism from $\mathrm{Frac}(B)$ to $\mathrm{Pic}(B)$ which, since principal ideals are free modules, factors through say $\alpha : \mathrm{ICG}(B) \longrightarrow \mathrm{Pic}(B)$.

Conversely, if M is a locally-free B-module of rank 1 ,   then tensoring the inclusion $B \hookrightarrow K$ with M we obtain a monomorphism

$$M = B \otimes_B M \longrightarrow K \otimes_B M .$$

Now $K \otimes_B M$ is isomorphic to K as a K-module;   choose an isomorphism (which is well-defined up to K^*) and we can identify M with a sub-B-module of K , which is in fact a fractional ideal, so determines an element $\beta(M)$ of $\mathrm{ICG}(B)$.   The map $\beta : \mathrm{Pic}(B) \longrightarrow \mathrm{ICG}(B)$ is clearly inverse to α .

Suppose now that $\mathrm{char}\, K = p \neq 0$.   The elements of $\mathrm{Fr}(P)$ are called __divisors__ on K. Let $\deg : Fr(P) \to \mathbb{Z}$ be the homomorphism induced by the function $S \longmapsto \deg(S) = \deg(K(S)/k)$ and let G_0 denote its kernel. By the Artin-Whaples formula 4.7.5 the image of the

map $\phi : K \longrightarrow Fr(P_0)$, $x \longmapsto \sum_S \mathrm{ord}_S(x)S$ which is called the group of __principal divisors__ and is denoted by G_1, is contained in G_0.

__Definition 5.4.3__   The group G_0/G_1 is called the __divisor class group__ of K and is denoted by D_0.

The group D_0 depends on the field K, and not the extension K/F.

__Theorem 5.4.4__   *The divisor class group of a global field of characteristic non-zero is finite.*

The proof of this is sketched in the appendix §5.6 in terms of idèles, and is essentially due to Riemann.   We now show the relation between D_0 and $ICG(B)$.

Let $h : G_0 \to Frac(B)$ denote the composite

$$G_0 \hookrightarrow Fr(P) \xrightarrow{\ j\ } Fr(P_0) = Frac(P)$$

where j is induced by the decomposition $P = P_0 \sqcup P_\infty$. The group $\mathrm{Ker}\, h$ is a free abelian group of rank $r - 1$. Let W, X and Y be the groups for which the following diagram (which has exact rows) has exact columns:

__5.4.5__

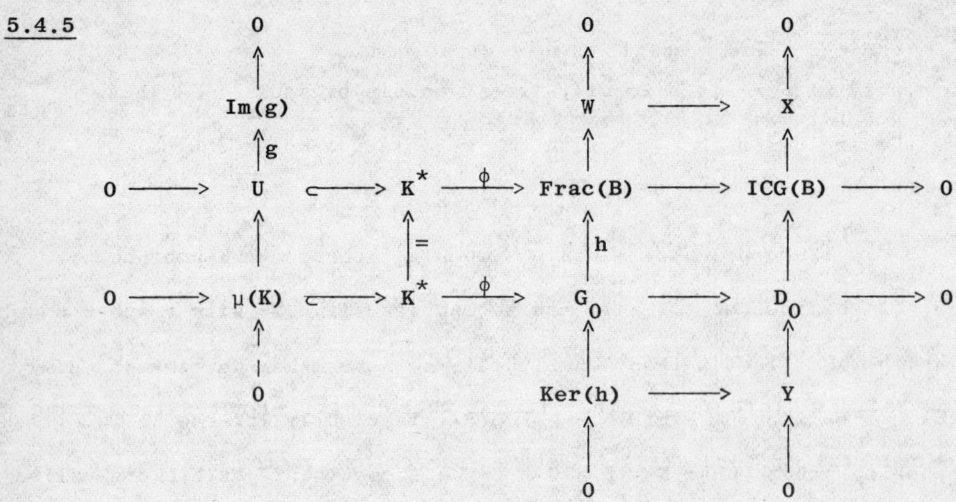

__Proposition 5.4.6__   *Choose* $g' : \mathrm{Im}(g) \to U$ *such that* $g'g = 1$, *and put*

$s = \phi g'$.

(5.4.6.1)   *The map* $W \to X$ *is an isomorphism.*

(5.4.6.2)   *The sequence* $0 \to \mathrm{Im}(g) \overset{s}{\to} \mathrm{Ker}\ e \to Y \to 0$ *is exact.*

(5.4.6.3)   *The order of* Y *is* $Rf/(\overset{r}{\underset{i=1}{\Pi}} f_i)(\log q)^{r-1}$, *where* $f_i = \deg(\infty_i)$,

        $f = \mathrm{hcf}\{f_1,\ldots,f_r\}$ *and* R *is the regulator of* K/F.

(5.4.6.4)   *The group* W *is cyclic of order* f/d *where*

        $d = \mathrm{hcf}\{\deg(S) : S \in P\}$.

__Proof__   The first two parts are obtained by simple diagram chasing; the
fact that s is a monomorphism is a restatement of 5.3.1.   Since $\mathrm{Im}(g)$
and $\mathrm{Ker}(h)$ are free of the same rank, the order Y is equal to $|\det(s)|$.

There is a natural isomorphism $\mathrm{Ker}(h) \to \tilde{\mathrm{F}}\mathrm{r}(P_\infty)$ (where $\tilde{\mathrm{F}}\mathrm{r}(P_\infty)$ denotes
$\mathrm{Ker}(\deg : \mathrm{Fr}(P_\infty) \to \mathbb{Z}))$ such that the diagram

$$
\begin{array}{ccccc}
U & \xrightarrow{\ \ \mathrm{ord}\ \ } & \tilde{\mathrm{F}}\mathrm{r}(P\ ) & \hookrightarrow & \mathrm{Fr}(P_\infty) \\[4pt]
g'\uparrow & & \uparrow & & \\[4pt]
\mathrm{Im}(g) & \xrightarrow{\ \ s\ \ } & \mathrm{Ker}(h) & &
\end{array}
$$

commutes, where $\mathrm{ord}(x) = \overset{r}{\underset{i=1}{\Sigma}} \mathrm{ord}_{\infty_i}(x)\infty_i$.   Choose $a_1,\ldots,a_r \in \mathbb{Z}$ such
that $f = \overset{r}{\underset{i=1}{\Sigma}} a_i f_i$ and let $d' : \mathbb{Z} \to \mathrm{Fr}(P_\infty)$ be the homomorphism such that
$d'(1) = \underset{i=1}{\Sigma} a_i \infty_i$,   which splits the exact sequence

$$
0 \longrightarrow \tilde{\mathrm{F}}\mathrm{r}(P_\infty) \longrightarrow \tilde{\mathrm{F}}\mathrm{r}(P_\infty) \xrightarrow{\ \deg/f\ } \mathbb{Z} \longrightarrow 0.
$$

If s' denotes the composite $\mathrm{Im}(g) \oplus \mathbb{Z} \longrightarrow \tilde{\mathrm{F}}\mathrm{r}(P_\infty) \oplus \mathbb{Z} \xrightarrow{\ (\mathrm{inc},d')\ } \mathrm{Fr}(P_\infty)$
then clearly $|\det(s)| = |\det(\mathrm{ord})| = |\det(s')|$.   If $\{e_1,\ldots,e_{r-1}\}$ is a
fundamental set of units for U, then with respect to the basis
$\{g(e_1),\ldots,g(e_{r-1})\}$ for $\mathrm{Im}(g)$ and the basis $\{\infty_1,\ldots,\infty_r\}$ for $\mathrm{Fr}(P_\infty)$ the
matrix of s' is $(o(e_1),\ldots,o(e_{r-1}),a)$, where $o(x)$ denotes the column
vector $(\mathrm{ord}_{\infty_1}(x),\ldots,\mathrm{ord}_{\infty_r}(x))^t$ and a the column vector $(a_1,\ldots,a_r)^t$.

The determinant of this matrix has absolute value $Rf / (\prod_{i=1}^{r} f_i) (\log q)^{r-1}$ by 5.3.4.

Finally, more diagram chasing shows that

$$W \stackrel{\sim}{=} \mathrm{Im}(\deg : \mathrm{Fr}(P) \to \mathbb{Z}) / \mathrm{Im}(\deg : \mathrm{Fr}(P_\infty) \to \mathbb{Z}) .$$

<u>Remark 5.4.7</u>   It can be shown[68] that the integer d is precisely $\deg(k/F_p)$, i.e. $\log_p q$.

<u>Corollary 5.4.9</u>   *The ideal class group* $\mathrm{ICG}(B)$ *is finite of order*

$$\frac{(\prod_{i=1}^{r} f_i) (\log q)^{r-1} . |D_0|}{R . \deg(k/F_p)} .$$

If on the other hand $\mathrm{char}\, K = 0$, there is no algebraic analogue of the map 'deg' or of the groups G_0 , G_1 or D_0. The presence of non-discrete spots means that we must take into account the topology of the situation, which is no longer discrete, and this is one reason for introducing adèles. The finiteness of $\mathrm{ICG}(B)$ in this case will be discussed in §5.6 .

5.5   THE ζ-FUNCTIONS OF B AND K

For any non-zero ideal I of B, the quotient ring B/I is finite of order $N(I) = \prod_{S \in P_0} q_S^{\mathrm{ord}_S(I)}$.

<u>Theorem 5.5.1</u>   *The infinite products*

$$\zeta(z, K/F) = \prod_{S \in P_0} (1 - q_S^{-z})^{-1} \quad and \quad \zeta(z, K) = \prod_{S \in P_{na}} (1 - q_S^{-z})^{-1}$$

converge absolutely for $\mathrm{Re}(z) > 1$. *Moreover for any integer* m *the set* $\{I \lhd B : N(I) = m\}$ *is finite and the Dirichlet series*

$$\sum_{\substack{I \lhd B \\ I \neq 0}} N(I)^{-z} = \sum_{m} \frac{|\{I \lhd B : N(I) = m\}|}{m^z}$$

also converges absolutely for $\mathrm{Re}(z) > 1$, *and converges to* $\zeta(z, K/F)$.

<u>Proof</u>   It suffices to prove absolute convergence when z is real. For any spot $S \in P_{na}(F)$ the subset $S^K \subset P_{na}(K)$ has at most n elements (see

4.2.3).   Hence we have

$$\prod_{T \in S^K} (1 - q_T^{-z})^{-1} \le (1 - q_S^{-z})^{-n} \quad \text{for} \quad z \text{ real,} \quad z > 0$$

since $q_T \ge q_S$.   The infinite products $\zeta(z,K)$ and $\zeta(z,K/F)$ thus

converge absolutely for $z \ge 1$ by comparison with the infinite products for

$\zeta(z,F)^n$ and $\zeta(z,F/F)^n$.   Similar remarks apply to the Dirichlet series.

Now expanding the product $\zeta(z,K/F)$ as in §1.3, and using unique

factorisation 5.2.8 and 5.2.9.3 one easily shows equality between $\zeta(z,K/F)$

and the Dirichlet series $\sum\limits_{\substack{I \lhd B \\ I \neq 0}} N(I)^{-z}$.

Hence the zeta functions $\zeta(z,K)$ and $\zeta(z,K/F)$ are holomorphic in H_1 .

We note that if char $K = 0$, then $\zeta(z,K) = \zeta(z,K/F)$, but if char $K = p \neq 0$

then we have $\zeta(z,K) = \left[\prod\limits_{i=1}^{r} (1 - q_{\infty_i}^{-z})^{-1} \right] \zeta(z,K/F)$.   Moreover in the latter

case, since $q_S = q^{\deg(S)}$ for any $S \in P(K)$, the infinite products

5.5.2 $\qquad\qquad Z(t,K/F) = \prod\limits_{S \in P_0} (1 - t^{\deg(S)})^{-1}$

and

$$Z(t,K) = \left[\prod\limits_{i=1}^{r} (1 - t^{\deg(\infty_i)})^{-1} \right] Z(t,K/F)$$

are absolutely convergent in the disc $\{t \in C : |t| < 1/q\}$ and we have

$\zeta(z,K) = Z(q^{-z},K)$ and $\zeta(z,K/F) = Z(q^{-z},K/F)$.

If we define for any non-zero ideal I of B the integer $\deg(I)$

(called the <u>degree</u> of I) to be $\log_q N(I)$, so that $\deg(p_S) = \deg(S)$, then

it follows from 5.5.1 and 5.5.2 that the power series $\sum\limits_{\substack{I \lhd B \\ I \neq 0}} t^{\deg(I)}$ converges

absolutely in the disc $\{t \in C : |t| < 1/q\}$ and converges to $Z(t,K/F)$.

By techniques of Fourier analysis over the locally – compact fields K_S

for $S \in P(K)$ which is beyond the scope of these notes but which we sketch

in §5.6, one can prove various analytic properties of $\zeta(z,K)$ which if

char $K = 0$ contain all the results of §2.3 and §2.4, and which if K is a

quadratic extension of $F_p(X)$ contain the results of §3.2 and §3.3.

This then achieves the objective of this chapter, namely to present algebraic theory in terms of the analytic theory of valuations.

Rather than repeat the results of the classical case in this context, we state only the interpretation of these results in the case of an algebraic function field.

Theorem 5.5.3   *Let* K *be a global field of characteristic* $p \neq 0$, *and field of constants of order* q .   *The holomorphic function* Z(t,K) *in the disc* $\{t \in C : |t| < 1/q\}$ *extends to a meromorphic function on the complex plane, with poles at* 1 *and* 1/q , *both of which are simple.   Moreover, this function satisfies the functional equation*

$$Z(1/qt,K) \;=\; (qt^2)^{1-g} Z(t,K)$$

where g *is the genus of* K .   *The residue of* Z(t,K) *at* $t = 1/q$ *is* $|D_0|/q^g(1-q)$, *where* D_0 *is the divisor class group of* K .

Corollary 5.5.4   *Under the conditions of 5.5.3 the residue of* $\zeta(z,K/F)$ *at* $z = 1$ *is*

$$\frac{Rfhq^{1-g}}{(1-q)(\log q)^{r-1}} \; \prod_{i=1}^{r} \left(\frac{1 - q^{-f_i}}{f_i} \right)$$

where h *is the class number of* K/F .

Proof   This is immediate since $\zeta(z,K/F) = Z(p^{-z},K) \prod_{i=1}^{r} (1 - q_{\infty_i}^{-z})$ and because of the relation 5.4.8 between $|D_0|$ and h .

5.6   APPENDIX: ADÈLES AND IDÈLES

Let K be a global field.   Consider the topological ring $\tilde{K} = \prod_{S \in P} K_S$. If a is an element of $\tilde{K}$, we write a_S for its component in K_S .   By 4.7.5 the image of the natural map $K \to \tilde{K}$ is contained in the subring $A_K = \{x \in \tilde{K} : x_S \in R_S$ for almost all $S \in P_{na}\}$.   This ring is called the **ring of adèles** (or valuation vectors) of K .   It is the smallest subring of K that contains K_S for each $S \in P$ as well as the ring $\prod_{S \in P_{na}} R_S$.   The group of units of A_K is called the idèle group of K, and is denoted by J or J_K. Thus we have

$$J_K = \{x \in \tilde{K} : x_S \neq 0 \text{ for all } S \in P \text{ and } x_S \in U_S \text{ for almost all } S \in P_{na}\}.$$

These two algebraic objects carry a natural locally-compact topology compatible with the algebraic structure, which we now describe.

First we note that as a subspace of $\tilde{K}$, the set A_K is not locally-compact, no more than $\tilde{K}$ is. For any finite set $X \subset P$ which contains P_a, let $A_K(X) = \{x \in A_K : x_S \in R_S \text{ for all } S \in P \setminus X\}$. Such finite subsets are directed by inclusion, and clearly as rings

$$A_K = \varinjlim_X A_K(X).$$

Now $A_K(X)$ as a subspace of $\tilde{K}$ is locally-compact, being a product of a finite number of locally-compact fields and a product of compact rings. One gives A_K the direct limit topology, so that $A_K = \varinjlim_X A_K(X)$ is an isomorphism of topological rings, and hence A_K is locally-compact. Equivalently the topology on A_K is the smallest satisfying the conditions that A_K is a topological ring and the induced topology on $(\prod_{S \in P_a} K_S) \times (\prod_{S \in P_{na}} R_S)$ is the product topology.

This is an example of a restricted product topology[69].

The subset J_K of A_K with the subspace topology is not a topological group since the map $x \mapsto x^{-1}$ is not continuous. There are three equivalent ways of defining the topology on J_K. Firstly by analogy with the above if X is a finite subset of P containing P_a, let $J_K(X) = J_K \cap A_K(X)$, so that $J_K = \varinjlim_X J_K(X)$ as groups, and give J_K the direct limit topology.

Secondly give J_K the subspace topology induced from the embedding $J_K \to A_K \times A_K, \ x \mapsto (x, x^{-1}).$

Thirdly, the group $Aut(A_K)$ of (continuous) additive automorphisms of A_K with the compact-open topology is a locally-compact group[70], and one gives J_K the subspace topology induced from the embedding $J_K \to Aut(A_K)$.

It is not the intention of these notes to describe in any detail the theory of adèles and idèles but it is important to emphasise the rôle of harmonic analysis.   We suffice to list important properties which are needed[71].

<u>5.6.1</u>   *The topologies induced on* K *and* K^* *by the embeddings in* A_K *and* J_K *are discrete.*

<u>5.6.2</u>   *The map* $|.| : A_K \to R$ *defined by* $|x| = \prod_{S \in P} |x_S|_S$ *(which is called the norm map) is continuous.*   Its restriction to $J_K \subset A_K$ is also continuous (where J_K of course has the topology described above, not the subspace topology) and is homomorphism into R^* surjective if char $K = 0$, but having discrete image if char $K \neq 0$.   We define $J^1 = J^1_K$ to be the kernel of this map, and M to be its image, which is free Abelian of rank 1.

<u>5.6.3</u>   *The sequence* $0 \to K \to A_K \to A_K/K \to 0$ *is self-dual with respect to Pontrjagin duality.*   (This follows from the facts that each of the sequences $0 \to R_S \to K_S \to K_S/R_S \to 0$ for $S \in P_{na}$ and $0 \to Z \to R \to R/Z \to 0$ are self-dual, along with results on duals of restricted products[72].)

Now since A_K/K is the dual of the discrete group K , it is compact, but J_K/K^* is not compact, since it has quotient M .   However, we have the following theorem, essentially due to Riemann[73].

<u>Theorem 5.6.4</u>   *The group* J^1/K^* *is compact.*

The proof[74] is a simple application of the Strong Approximation Theorem 5.2.4, and is essentially the step referred to in 5.3.2 for showing that $Im(g) \subset H$ is a lattice.

The following proof that the ζ-function extends to a meromorphic function in C and satisfies the functional equation is the analogue of the standard proof for the ζ-function $\zeta(z,Q)$ in terms of the theta function and its functional equation[75].

Let α be the Haar measure on A_K such that $\alpha(A_K/K) = 1$. Let ν be the Haar measure on J such that on the quotient $J/K^* = J^1/K^* \times M$ the induced measure (also denoted by ν) satisfies the two conditions that $\nu(J^1/K^*) = 1$, and on M we have $d\nu(x) = \frac{dx}{x}$ if $\operatorname{char} K = 0$, $\nu(\{x\}) = 1$ if $\operatorname{char} K \neq 0$. The measures α and ν induce Haar measures α_S and ν_S on K_S and K_S^* such that $\alpha = \Pi\alpha_S$ and $\nu = \Pi\nu_S$.

Let $\Phi_S : K_S \to R$ be the continuous function defined by

$$\Phi_S(x) = \exp\left(-\pi \, |x|_S^2\right) \quad \text{if} \quad S \text{ is a real spot}$$

$$= \exp\left(-2\pi \, |x|_S\right) \quad \text{if} \quad S \text{ is a complex spot}$$

$$= \left\{ \begin{array}{ll} 1 & \text{if } x \in R_S \\ 0 & \text{if } x \notin R_S \end{array} \right\} \quad \text{if} \quad S \text{ is non-archimidean.}$$

The function $\Phi : A_K \to R$, $x \mapsto \underset{S}{\Pi} \Phi_S(x)$ is continuous.

<u>Notation 5.6.5</u> For any function $f : A_K \to C$ and $z \in C$ let $I(z,f)$ denote the integral $\displaystyle\int_J f(x) |x|^z \, d\nu$.

<u>5.6.6</u> Now formally we have $\displaystyle I(z,\Phi) = \int_J \underset{S}{\Pi} \Phi_S(x_S) |x_S|_S^z \, d\nu$

$$= \underset{S}{\Pi} \left(\int_{K_S^*} \Phi_S(x_S) |x_S|_S^z \, d\nu_S \right).$$

<u>Notation 5.6.7</u> For two complex valued functions f, g defined on a set X, we write $f \doteq g$ if there is an $r \in R^*$ such that $f(x) = rg(x)$ for all $x \in X$.

<u>Lemma 5.6.8</u> *We have the following for* $\operatorname{Re} z > 0$:

$$\int_{K_S^*} \Phi(t) |t|_S^z \, d\nu_S \; \doteq \; \pi^{z/2} \Gamma(z/2) \quad \textit{if } S \textit{ is a real spot}$$

$$\doteq \; (2\pi)^{1-z} \Gamma(z) \quad \textit{if } S \textit{ is a complex spot}$$

$$\doteq \; (1 - q_S^{-z})^{-1} \quad \textit{if } S \textit{ is non-archimidean.}$$

__Proof__   For the real or complex spot, the result is obvious.   Suppose S is non-archimidean.   Let $U_n = m_S^n \setminus m_S^{n+1}$ so that for $t \in U_n$ we have $|t| = q_S^{-n}$.   Since $d\alpha_S / |t|$ is a Haar measure on K_S^* , we have

$$\int_{K_S^*} \Phi_S(t) |t|_S^z d\nu_S = \int_{R_S \setminus \{0\}} |t|_S^{z-1} d\alpha_S = \sum_{n=0}^{\infty} \int_{U_n} |t|_S^{z-1} d\alpha_S = \sum_{n=0}^{\infty} q_S^{-n(z-1)} \alpha(U_n) .$$

Since $\alpha(m_S^n) = q_S^{-n} \alpha(R)$ (see 4.8.7), we see that $\alpha(U_n) = \alpha(R)(q_S^{-n} - q_S^{-n-1})$. Substituting this value of $\alpha(U_n)$ and summing the resulting geometric series, we obtain the required result.

Hence formally we have

$$I(z,\Phi) \doteq (\pi^{-z/2} \Gamma(z/2))^{r_1} ((2\pi)^{1-z} \Gamma(z))^{r_2} \zeta(z,K) \quad \text{if char } K = 0$$

$$\doteq \zeta(z,K) \quad \text{if char } K \neq 0 .$$

But now the absolute convergence of the infinite product (5.6.6) for $\text{Re } z > 1$ (which follows from the previous equation and 5.5.1) implies the absolute convergence of the integral $I(z,\Phi)$ for $\text{Re } z > 1$.

Let $F_0 : J \to R$ denote the function which takes the value $0 , \tfrac{1}{2}$ or 1 according as $|x| > 1$, $|x| = 1$ or $|x| > 1$. Let $F_1(x) = F_0(x^{-1})$.   Put, for any $f : A_K \to C$,

$$I_i(z,f) = \int_J f(x) |x|^z F_i(x) d\nu \quad \text{for} \quad i = 0 , 1$$

so that $I(z,\Phi) = I_0(z,\Phi) + I_1(z,\Phi)$.   It follows from previous remarks that $I_i(z,\Phi)$ is absolutely convergent for $\text{Re } z > 1$.

Now if $r,s \in R$ are such that $s > 1$ and $s > r$, then $|x|^r \leq |x|^s$ if $|x| \geq 1$, and so the integral $I_0(r,\Phi)$ is majorised by the integral $I_0(s,\Phi)$, and so $I_0(z,\Phi)$ is absolutely convergent for all $z \in C$.

__Lemma 5.6.9__   *The function* $I_0(z,\Phi)$ *is holomorphic in the complex plane.*

__Proof__   If δ , t are real numbers such that $\delta > 0 , t > 1$ then for any $z , z' \in C$ with $0 < |z - z'| < \delta$ we have

$$\left| \frac{t^z - t^{z'}}{z - z'} \right| \leq t^{|z'| + \delta} \log t \leq t^{|z'| + \delta + 1}$$

and since the function $\Phi(x) |x|^{|z'| + \delta + 1} F_0(x)$ is integrable, it follows from the Dominated Convergence Theorem[76] that $I_0'(z', \Phi)$ exists and is equal to

$$\int_J \Phi(x) |x|^z \log |x| F_0(x) d\nu .$$

For every $S \in P(K)$ there is a non-trivial 'basic character' $e_S : K_S \to S^1$ with respect to which the function Φ_S is its own Fourier transform[77]. In particular if S is non-archimidean, this character is trivial on R_S.

Now choose a non-trivial character $\chi : A_K \to S^1$ which is trivial on K. This character induces a character $\chi_S : K_S \to S^1$ such that $\chi = \Pi \chi_S$, and moreover there is a unique element $a_S \in K_S$ such that $\chi_S(x) = e_S(a_S x)$ for all $x \in K_S$, and $a_S \neq 0$ since χ_S can be shown to be nontrivial[78]. The following result shows that harmonic analysis captures the algebra[79].

<u>Proposition 5.6.10</u> *The* element $a = (a_S) \in \tilde{K}$ *is an idéle, and if* char $K = 0$ *then* $|a| = |D|^{-1}$ *where* D *is the discriminant of* K *while if* char $K \neq 0$ *then* $|a| = q^{2-2g}$ *where* $q = |k|$ *and* g *is the genus of* K.

Let $\hat{\Phi} : A_K \to C$ be the Fourier transform of Φ with respect to the character χ. Since $\Phi(x) = \Pi_S \Phi_S(x_S)$, we have $\hat{\Phi}(x) = \Pi_S \hat{\Phi}_S(x_S)$, where $\hat{\Phi}_S$ is the Fourier transform of Φ_S with respect to the character χ_S on K_S. Now Φ_S is its own transform with respect to the character e_S and so it follows that

$$\hat{\Phi}_S(t) = |a_S|_S^{\frac{1}{2}} \Phi(a_S t) \quad \text{for} \quad t \in K_S ,$$

and hence

<u>5.6.11</u> $\qquad\qquad \hat{\Phi}(x) = |a|^{\frac{1}{2}} \Phi(ax) \quad \text{for} \quad x \in A_K .$

It follows from this equation and a modification of 5.6.9 that the integral $I_0(z, \hat{\Phi})$ is absolutely convergent and defines a holomorphic function in the complex plane.   The next theorem contains various results of Chapters 2 and 5 as special cases.

<u>Theorem 5.6.12</u>   *For* $\mathrm{Re}\ z > 1$ *we have*

$$I(z,\Phi) = I_0(z,\Phi) + I_0(1 - z,\hat{\Phi}) + \hat{\Phi}(0)\lambda(z - 1) - \hat{\Phi}(0)\lambda(z)$$

where $\lambda(z)$ *is a meromorphic function satisfying* $\lambda(z) = -\lambda(-z)$. *In particular, the function* $z \mapsto I(z,\Phi)$ *extends to a meromorphic function on the complex plane which satisfies the functional equation*

$$I(z,\Phi) = |a|^{-z+\frac{1}{2}} I(1 - z,\Phi) \quad \text{for} \quad z \in C .$$

<u>Proof</u>   Let z be a complex number such that $\mathrm{Re}\ z > 1$ consider the integral

(5.6.12.1)
$$I_1(z,\Phi) = \int_J \Phi(x)|x|^z F_1(x)d\nu = \int_{J/K^*} |x|^z F_1(x) \sum_{y \in K^*} \Phi(xy)d\nu .$$

Now the function $\Phi^x : A_K \to C$ $(y \to \Phi(xy))$ has Fourier transform given by

$$\hat{\Phi}^x(y) = \int_A \Phi(xu)\chi(uy)d\alpha(u) = \int_A \Phi(v)\chi(vx^{-1}y)|x|^{-1}d\alpha(v) \quad (\text{putting}\ u = x^{-1}v) ,$$

so that

(5.6.12.2)
$$\hat{\Phi}^x(y) = |x|^{-1}\hat{\Phi}(x^{-1}y) .$$

By virtue of the self-duality of the sequence $0 \to K \to A_K \to A_K/K \to 0$ we have the <u>Poisson Summation Formula</u>[80]

(5.6.12.3)
$$\sum_{y \in K} \Phi^x(y) = \sum_{y \in K} \hat{\Phi}^x(y) .$$

Putting together (5.6.12.2) and (5.6.12.1) we obtain the equation

(5.6.12.4)
$$\sum_{y \in K^*} \Phi(xy) = |x|^{-1}\{\hat{\Phi}(0) + \sum_{y \in K^*} \Phi(x^{-1}y)\} - \Phi(0)$$

and substituting (5.6.12.4) into (5.6.12.1) we see that

$$I_1(z,\Phi) = \int_{J/K^*} (\hat{\Phi}(0)|x|^{z-1} - \Phi(0)|x|^z)F_1(x)d\nu + \int_J |x|^{z-1}\hat{\Phi}(x^{-1})F_1(x)d\nu$$

$$= \hat{\Phi}(0)\lambda(z - 1) - \Phi(0)\lambda(z) + \int_J |x|^{1-z}\hat{\Phi}(x)F_0(x)d\nu \quad (\text{changing}\ x\ \text{to}\ x^{-1})$$

where $\lambda(z) = \int_{J/K^*} |x|^z F_1(x)d\upsilon ,$ that is to say

(5.6.12.5)
$$I_1(z,\Phi) = \hat{\Phi}(0)\lambda(z - 1) - \Phi(0)\lambda(z) + I_0(z,\hat{\Phi}) .$$

Now if char $K = 0$, then $\lambda(z) = \displaystyle\int_0^1 t^z \frac{dt}{t} = \frac{1}{z}$, but if char $K \neq 0$,

$$\lambda(z) = \frac{1}{2} + \sum_{n=1}^{\infty} Q^{-nz} = \frac{1}{2}\left(\frac{1 + Q^{-z}}{1 - Q^{-z}}\right) \quad \text{where } Q \in Z \text{ generates the group } M .$$

This proves the first part of the theorem.

Now similar remarks apply to the integral $I(z,\hat{\Phi})$ and so we obtain

(5.6.12.6) $I(z,\hat{\Phi}) = I_0(z,\hat{\Phi}) + I_0(1 - z,\hat{\hat{\Phi}}) + \hat{\hat{\Phi}}(0)\lambda(z - 1) - \hat{\Phi}(0)\lambda(z)$

and so using the Fourier inversion formula $(\hat{\hat{\Phi}}(x) = \Phi(-x))$ and the equation

$\lambda(z) = -\lambda(-z)$ we see that

(5.6.12.7) $I(z,\Phi) = I(1 - z,\hat{\Phi})$.

But we have $I(z,\hat{\Phi}) = \displaystyle\int_J \hat{\Phi}(x)|x|^z d\nu$

$$= \int_J \Phi(ax)|a|^{\frac{1}{2}}|x|^z d\nu = \int_J \Phi(ax)|ax|^z|a|^{z-\frac{1}{2}} d\nu$$

$$= |a|^{z-\frac{1}{2}} I(z,\Phi) ,$$

hence $I(z,\Phi) = I(1 - z,\hat{\Phi}) = |a|^{-z+\frac{1}{2}} I(1 - z,\Phi)$ and this the theorem is proved.

Next we prove finiteness of class number using the notation introduced

in §5.1.

For any finite set $X \subset P(K)$ containing P_a , let $J^1(X) = J(X) \cap J^1$.

Let $\sigma' : J^1 \to \mathrm{Frac}(B)$ be the homomorphism

$$\sigma'(x) = \sum_{S \in P_0} \mathrm{ord}_S(x_S) p_S$$

which is clearly surjective with kernel $J^1(P_\infty)$.   From the exact diagram

<u>5.6.13</u>

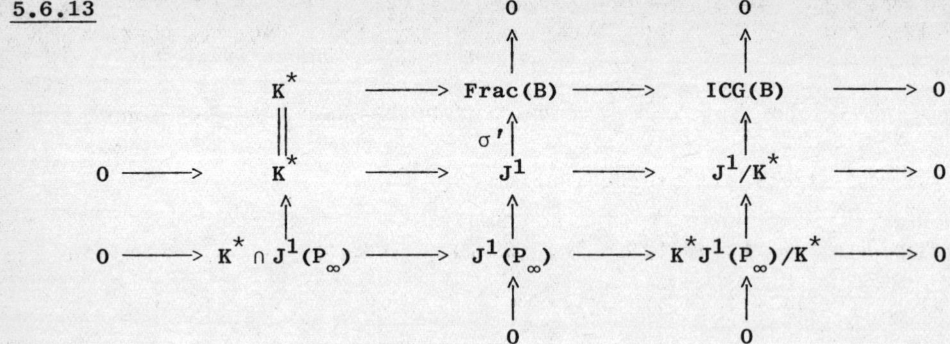

we deduce:

<u>Lemma 5.6.14</u>   *The groups ICG(B) and $J^1/K^*J^1(P_\infty)$ are isomorphic.*

In the case char (K) $\neq$ 0 ,   let $\sigma : J^1 \to G_0$ be the homomorphism

$$\sigma(x) = \sum_{S \in P} \mathrm{ord}_S(x_S).S .$$

From the commutative exact diagram

<u>5.6.15</u>

we deduce:

<u>Lemma 5.6.16</u>   *If* **char (K)** $\neq$ 0 ,   *then the divisor class group* D_0 *and the group* $J^1/K^*J^1(\phi)$ *are isomorphic.*

The finiteness of class group (see 5.4.5 and 5.4.9) now follow from 5.6.14, 5.6.15 and the next theorem.

<u>Theorem 5.6.17</u>   *The group* $J/K^*J(X)$ *is finite if* $X \neq \phi$.   *If* **char(K)** $\neq$ 0 *then* $J^1/K^*J^1(\phi)$ *is finite.*

<u>Proof</u>   In all cases $J(X)$ is open in J (by definition of the topology
as a direct limit) and so $K^*J^1(X)$ is open in J^1 , hence $J^1/K^*J^1(X)$ is
finite, being the quotient of the compact group J^1/K^* (see 5.6.4) by the
open subgroup $K^*J^1(X)/K^*$.

If $\mathrm{char}\,(K) = 0$ and X is a non-empty set of spots containing P_a ,
then the norm map $|\,.\,| : J(X) \to R_+^*$ is surjective.   It follows from the exact
diagram

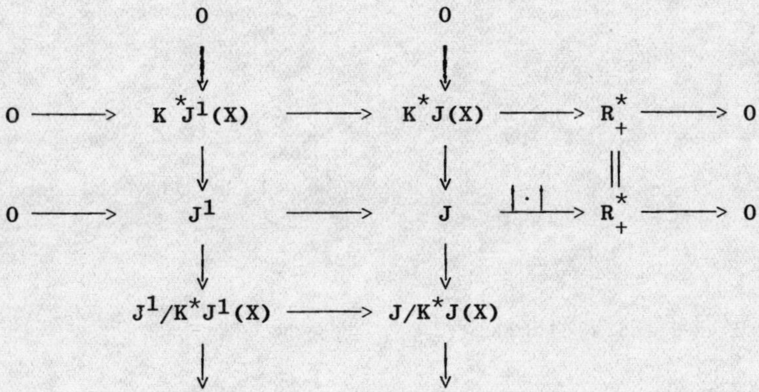

that $J/K^*J(X)$ is isomorphic to $J^1/K^*J^1(X)$, and so is finite.

If $\mathrm{char}\,(K) \neq 0$, then $J(\phi) = \prod_{S \in P} U_S = J^1(\phi)$.   Suppose that

$X \subset P(K)$ is non-empty, and consider the exact diagram.

$$
\begin{array}{ccccccccc}
 & & 0 & & 0 & & 0 & & \\
 & & \downarrow & & \downarrow & & \downarrow & & \\
0 \longrightarrow & \dfrac{K^*J(X) \cap J^1}{K^*J^1(\phi)} & \longrightarrow & \dfrac{J^1}{K^*J^1(\phi)} & \longrightarrow & \dfrac{J^1}{K^*J(X) \cap J^1} & \longrightarrow & 0 \\
 & \downarrow & & \downarrow & & \downarrow & & \\
0 \longrightarrow & \dfrac{K^*J(X)}{K^*J(\phi)} & \longrightarrow & \dfrac{J}{K^*J(\phi)} & \longrightarrow & \dfrac{J}{K^*J(X)} & \longrightarrow & 0 \\
 & \downarrow & & \downarrow{\scriptstyle |\cdot|} & & \downarrow & & \\
0 \longrightarrow & \dfrac{K^*J(X)}{K^*J(X) \cap J^1} & \longrightarrow & \dfrac{J}{J^1} & \longrightarrow & \dfrac{J}{J^1J(X)} & \longrightarrow & 0 \\
 & \downarrow & & \downarrow & & \downarrow & & \\
 & & 0 & & 0 & & 0 & &
\end{array}
$$

We have already shown that $J^1/K^*J^1(\phi)$ is finite, and so $J^1/K^*J(X) \cap J^1$

is finite. Since $J(X)$ is not contained in J^1, the group $K^* J(X) \cap J^1$ is a proper subgroup of $K^* J(X)$, so the quotient is non-trivial; but since the group $J/J^1 = M$ is free abelian of rank 1 this implies that the quotient group $J/J^1 J(X)$ is finite, and so $J/K^* J(X)$ is finite.

6 Algebraic curves

Throughout this chapter, k will denote a perfect field, with algebraic closure k^a/k and Galois group $G = \mathrm{Gal}(k^a/k)$.

6.1   PRELIMINARIES

We shall be interested in fields K with $\mathrm{tr.deg}(K/k) = 1$ in particular when k is finite (so that K is a global field) and when $k = \mathcal{C}$, which leads to the theory of (complex) Riemann surfaces.

As in §5.1, let P(K) denote the set of non-trivial spots on K which are trivial on k. Many of the results of Chapter 5 generalise to this situation, except possibly those which depend on finiteness of residue field. When we use such a generalisation, we shall refer to the relevant result with a letter 'e' appended, so that for example (4.7.5.1e) states that for $x \in K^*$, x belongs to U_T almost all $T \in P(K)$.

From 4.1.9 and 4.2.3 we know that all the spots in P(K) are discrete.

<u>Lemma 6.1.1</u>   *For any* $S \in P(K)$ *the extension* $K(S)/k$ *is finite.*

<u>Proof</u>   If X is an element of K such that K is a finite extension of $F = k(X)$, and T is the restriction of S to F, then $F(T)/k$ is finite by 4.1.10, so $K(S)/k$ is finite.

The completed field K_S is by 4.1.10 an extension[81] of a field of Laurent series. For any field F, the ring $F[[X]]$ of power series over F is a local domain with maximal ideal generated by X. The X-adic spot on $F((X))$ is complete and discrete, and has valuation ring $F[[X]]$ and residue

field F.   The reduction map $F[[X]] \to F$ associates to the power series
$a = \Sigma a_n X^n$ its constant term a_0 and we refer to this as 'substituting
$X = 0$ ' and write $a \mapsto a(0)$.   Henceforth in any reference to a spot on a
field $F((X))$ of Laurent series, it will be tacitly assumed that the spot is
the X-adic one.

6.2   PLANE CURVES

For any polynomial $g \in k[X,Y]$, let $V(g) = \{(x,y) \in A^2(k^a) : g(x,y) = 0\}$,
where for any field F, $A^n(F)$ denotes the n-dimensional affine space F^n.
The set $V(g)$ is infinite, and if $g \neq 0$ it is called a <u>plane curve</u>
or more precisely an affine k^a/k curve.   The projection $(x,y) \mapsto x$ from
the curve $V(g)$ to k^a is surjective, and the inverse image of the point
$x \in k^a$ is finite, except when the minimum polynomial of x over k divides
g, so that both $V(g)$ and its complement are infinite.   Both are invariant
under G.

     If $f \in k[X,Y]$ is irreducible and defines the curve $V = V(f)$, we say
that V is <u>irreducible</u>; in this case the quotient ring $k[V] = k[X,Y]/\langle f \rangle$
is an integral domain, and the elements of its field of fractions $k(V)$ are
called <u>rational functions</u> on V.   Since $k(V)$ is isomorphic to
$k(X)[Y]/\langle f \rangle$, it is a finite extension of $k(X)$ (and similarly of $k(Y)$),
separable if and only if f is separable as a polynomial in Y over $k(X)$.
Any field E of transcendence degree 1 over k is isomorphic[82] to $k(V)$
for some irreducible curve V determined by suitable choice of elements
$X,Y \in E$.   Initially, we shall concentrate on a particular extension
$k(V)/k(X)$ in relation to the geometry of V but in sections 6.6, 6.7, 6.8
we shall discuss the dependence on the choice of X and Y.

74

Lemma 6.2.1 *If* $f, g \in k[X,Y]$ *are non-zero polynomials with* f *irreducible and* $V(f) \cap V(g)$ *infinite, then* f *divides* g *and* $V(f) \subset V(g)$.

Proof If not, then the elements f, g of the euclidean domain $k(X)[Y]$ are relatively prime, so there are non-zero elements $a, b \in k[X,Y]$ and $c \in k[X]$ such that $af + bg = c$. But since f and g have infinitely many common zeros, c has infinitely many zeros, which is impossible.

Now suppose that $f \in k[X,Y]$ is irreducible, and is separable in Y. Put $V = V(f)$, $K = k(V)$, and $P(f) = \{S \in P(K) : X, Y \in R_S\}$. By (4.7.5.1e) the set $P(K) \setminus P(f)$ is finite, and if f is monic, consists of those spots whose restriction to $k(X)$ is the infinite spot. By 6.1.1 we can choose an embedding of $K(S)$ into k^a; let $(x,y) \in A^2(k^a)$ be the image of $(X,Y) \in A^2(K)$ under the composite $R_S \longrightarrow K(S) \longrightarrow k^a$. We say that S lies over (x,y). Since $f(X,Y) = 0$ in R_S, the point $P = (x,y)$ lies on V. A different choice of embedding $K(S) \longrightarrow k^a$ produces a point in the G-orbit of P, so there is a well-defined function $c : P(f) \longrightarrow V/G$ mapping S to the G-orbit of P, which we denote by $[P]$. If S lies over (x,y), then $x \in k^a$ represents the restriction of S to $k(X)$ (cf. 4.1.11).

Conversely, we shall show that for any $P = (x,y) \in V$ there is at least one spot $S \in P(f)$ lying over P. The element $x \in k^a$ defines a spot on $k(X)$ whose completion F is $k'((X-x))$, where $k' = k(x) \subset k^a$. Let $A \subset F$ be the valuation ring, namely $k'[[X-x]]$. Extensions of the spot x to K correspond (by 4.2.6) to the irreducible factors of f over F. By Gauss's Lemma, there is a factorisation $f = \prod f_i$ of f into irreducible factors over F with $f_i \in A[Y]$ for each i. Since $f(x,y) = 0$ there is at least one i for which $f_i(x,y) = 0$; put $g = f_i$ and let E be the splitting field of g over F. Let R be the valuation ring of the spot on E extending the (X-x)-adic one on F, m its maximal ideal and U = R\m its group of units.   If $Y_1, \ldots, Y_d$ are the roots of g in E, then the spot on K corresponding to g is induced from the embedding of K = k(X)[Y]/f into E sending Y to Y_1.

First we show that Y_1 is integral.   Assume for a contradiction that $\mathrm{ord}\, Y_1 = -r$, where $r \in N$.   Since all the roots of g are conjugate, $\mathrm{ord}\, Y_j = -r$ for each j.   Over E we have $g(Y) = a \prod (Y - Y_j)$ for some $a \in E$, but again we can invoke Gauss's Lemma and write $g(Y) = a' \prod (t^r Y - t^r Y_j)$ where $t \in R$ is a prime element, $a' \in R$ and $Z_j = t^r Y_j$ is in U.   Since g is irreducible over F, a' must lie in R. Applying to this equation the quotient map $\rho : R \to R/m$, we obtain $0 = g(x,y) = \rho(a') \prod (-\rho(Z_j)) \neq 0$, which is a contradiction, so $Y_1 \in R$ so is integral.   Up to a unit in A, g is monic.   So S belongs to P(f). Next we show that $c(S) = [P]$.   Consider the A-algebra homomorphism $r : A[Y] \to R$ taking Y to y_1, and the evaluation homomorphism $e : A[Y] \to k^a$, $g \mapsto g(x,y)$.   Since x and y are algebraic over k, the image of e is a subfield of k^a, so M = Ker(e) is a maximal ideal of A[Y], and moreover $r(M) \subset m$.   The embedding $e : A[Y]/M \to k^a$ extends[83] to an embedding $e : K(S) = R/m \to k^a$ taking X,Y to x,y, so indeed $c(S) = [P]$.   Thus we have proved the following result.

<u>Proposition 6.2.2</u>   *If* $f(X,Y) \in k[X,Y]$ *is an irreducible polynomial, separable and of degree* n *in* Y, *defining the curve* V = V(f) *and we put* K = k(V) ,   *then the function* $c : P(f) \to V/G$ *is surjective and the inverse image of each point contains at most* n *points.*

6.3   SMOOTHNESS

Suppose in the proof of 6.2.2 there were two distinct values i, j such that
$f_i(x,y) = f_j(x,y) = 0$.   Differentiating $f = \Pi\, f_i$ with respect to Y and
$X - x$, we see that $\partial f/\partial Y$ and $\partial f/\partial X$ both vanish at (x,y).

<u>Definition 6.3.1</u>   A point $P = (x,y)$ on the curve $V(g)$ defined by the
polynomial $g \in k[X,Y]$ is <u>singular</u> if $\partial f/\partial X$ and $\partial f/\partial Y$ both vanish at P,
otherwise it is called <u>non-singular</u> or <u>simple</u>.

The curve $V(g)$ is called <u>non-singular</u> or <u>smooth</u> if all its points are
simple.   We note that if $V(g)$ is smooth, then g is separable in either
X or Y.   If $f \in k[X,Y]$ is irreducible, separable in Y, then f does
not divide $\partial f/\partial Y$ so $V(f)$ has only finitely many singular points by 6.2.1.

<u>Corollary 6.3.2</u>   *Under the hypotheses of 6.2.2, for almost all points [P]
of V/G, there is a unique spot S $\in$ P(f) lying over P, the exceptions
being possibly the orbits of the singular points.   In particular, if V is
smooth, then c: P(f) $\longrightarrow$ V/G is bijective* .

<u>Example 6.3.3</u>   The irreducible curve defined by $f = Y^2 - X^3$ has a singular
point at $(0,0)$, but only one spot lies over this point since f is
irreducible over $k((X))$.

<u>Example 6.3.4</u>   If $\operatorname{char} k \neq 2,3$ then the curve V defined by
$f = X^3 + Y^3 + 3XY$ has $(0,0)$ as a singular point, but over $k((X))$ the
polynomial f has a linear and an irreducible quadratic factor, so there are
two spots lying over this point.

6.4   INTEGERS

Suppose that $f \in k[X,Y]$ is irreducible, and is monic and separable of degree n in Y, and defines the curve $V = V(f)$. Let B be the integral closure of $k[X]$ in $k[V] = k[X,Y]/\langle f \rangle$. For example, if k is finite, then B is the ring of integers in the global field $k(V)/k(X)$. We shall investigate the relation between B and $k[V]$.

Both these rings are free $k[X]$ modules, B by 5.1.1e and $k[V]$ since $\{1, Y, \ldots, Y^{n-1}\}$ is a basis. Clearly $k[V]$ is contained in B, but in general B is larger (for example if $f = Y^2 - X^3$ then $Y/X \in B$).

If $P = (x,y)$ is a point of V, then evaluation at P is a homomorphism $e_P: k[X,Y] \longrightarrow k^a$ whose image $k(P)$ is a field, since x and y are algebraic over k. In fact $k(P) = k(x,y)$. The map e_P factors through $k[V]$; let $m_P \in \max k[V]$ be its kernel. For $\sigma \in G$, $m_{\sigma(p)} = m_P$ so $[P] \longmapsto m_P$ is a well-defined function from V/G to $\max k[V]$.

Conversely, if $m \in \max k[V]$, let $M \in \max k[X,Y]$ denote its inverse image. The field $E = k[X,Y]/M$ is a finite extension of $F = k[X]/k[X] \cap M$, and since Y generates E over F subject to $f(X,Y) = 0$, the extension E/F is finite, so E/k is finite. Choose an embedding of E in k^a, and let x,y denote the images of X, Y under the composite $e : k[X,Y] \longrightarrow k[X,Y]/M = E \longrightarrow k^a$. The point $P = (x,y)$ lies on V and $e = e_P$. The following is now trivial.

<u>Proposition 6.4.1</u> *If $f \in k[X,Y]$ is irreducible, monic in Y, and defines the curve $V = V(f)$, then the function $[P] \longmapsto m_P$ from V/G to $\max k[V]$ is a bijection.*

Now the ring B is a Dedekind domain by (5.2.9.1e). The function which associated to $M \in \max B$ the M-adic spot is a bijection from $\max B$ to $P(f)$. We have a commutative diagram (where r is the restriction map, $M \longmapsto k[V] \cap M$)

$$P(f) \xrightarrow{\;\;c\;\;} V/G$$

$$\text{max } B \xrightarrow[\;\;r\;\;]{} \text{max } k[V]$$

If V is smooth, then c is a bijection by 6.3.2, so r is a bijection. In fact one can show more. If $P \in V$ corresponds to the spot $S \in P(f)$, then the ring R_S is precisely the localisation of $k[V]$ at the maximal ideal m_P. But R_S is also (by 4.3.6) the localisation of B at the maximal ideal $M \in \text{max } B$ corresponding to S, whose restriction to $k[V]$ is m_P. Since integral closure is a local property[84] the smoothness of V implies that $k[V]$ is integrally closed. The converse is also true, but more difficult to prove[85].

Theorem 6.4.3 *If* $f \in k[X,Y]$ *is an irreducible polynomial, monic in* Y, *defining the curve* $V = V(f)$, *then the ring* $k[V]$ *is integrally closed if and only if* V *is smooth.*

6.5   ZETA FUNCTIONS REVISITED

Let k be a finite field of order q. The field k^a contains for each $m \in N$ a unique subfield k_m of order q^m. The extension k_m/k is normal and has cyclic Galois group of order m, generated by the Frobenius map $\pi : x \mapsto x^q$. The field k_m is contained in $k_{m'}$ if and only if m divides m', so that the group $G = \text{Gal}(k^a/k)$ is (isomorphic to) $\hat{Z}$, the profinite completion of the integers $\varprojlim Z/nZ$, and is generated topologically by π.

Theorem 6.5.1 *Let* k *be a finite field and* $f \in k[X,Y]$ *an irreducible polynomial, monic and separable in* Y *such that the curve* $V = V(f)$ *defined* *by* f *is smooth, and let* $K = k(V) = k(X)[Y]/\langle f \rangle$. *If* ν_m *denotes the number of solutions of the equation* $f(x,y) = 0$ *in* k_m *then*

$$Z(t, K/k(X)) = \exp \Sigma \, \nu_m \, t^m / m.$$

<u>Proof.</u> Since f is monic in Y, the elements of $P(f)$ are precisely the spots in $P(k)$ whose restriction to $k(X)$ is not the infinite one. Now if the spot $S \in P(f)$ lies over $P \in V$, the residue field $K(S)$ is isomorphic to $k(P)$, and so the number of points in the G-orbit of P is $\deg S$. But

$$Z(t) = Z(t, K/k(X)) = \prod_{S \in P(f)} (1 - t^{\deg(S)})^{-1} \quad \text{and so}$$

$$t\frac{d}{dt} \log Z(t) = \sum_{S \in P(f)} \frac{\deg(S)\, t^{\deg(S)}}{1 - t^{\deg(S)}} = \sum_{S \in P(f)} \deg(S) \sum_{n=1}^{\infty} t^{n\deg(S)} = \sum_{m=1}^{\infty} \nu_m \, t^m.$$

6.6 BIRATIONAL EQUIVALENCE

Suppose K is a field with $\mathrm{tr.deg}(K/k) = 1$ and that for $i = 1,2$ there are elements $X_i, Y_i \in K$ and irreducible polynomials $f_i \subset k[X,Y]$, separable in Y, such that $K/k(X_i)$ is a finite extension generated by the element Y_i subject to the equation $f_i(X_i, Y_i) = 0$. Let V_i be the curve $V(f_i)$.

Since X_2, Y_2 belong to $K = k(X_1)[Y_1]/\langle f_1 \rangle$, there are rational functions $g(X,Y)$, $h(X,Y)$ such that $X_2 = g(X_1, Y_1)$ and $Y_2 = h(X_1, Y_1)$. Write $g = g_0/g_1$ and $h = h_0/h_1$ where $g_0, g_1, h_0, h_1 \in k[X,Y]$. The sets $V(g_1)$ and $V(h_1)$ meet $V(f_1)$ in a finite set of points by 6.2.1. Let $Z_1 = V(g_1) \cup V(h_1)$.

For $(x,y) \in V(f_1) \setminus Z_1$, the elements $g(x,y) = g_0(x,y)/g_1(x,y)$ and $h(x,y) = h_0(x,y)/h_1(x,y)$ of k^a are defined, and the point $(g(x,y), h(x,y))$ lies on $V(f_2)$, that is to say we have a function $r_1 : V(f_1) \setminus Z_1 \to V(f_2)$ such that the coordinates of $r_1(x,y)$ are rational functions of x and y.

80

Similarly, there is a function $r_2 : V(f_2) \setminus Z_2 \to V(f_1)$ where Z_2 is a finite set of points, and the composites $r_1 r_2$, $r_2 r_1$ are both the identity on their relevant domains of definition, which consist of all but a finite set of points of the appropriate curve.

<u>Definition 6.6.1</u> Two irreducible plane curves V_1, V_2 are <u>birationally</u> <u>equivalent</u> if there are finite subsets Z_i of V_i and functions $r_1 : V_1 \setminus Z_1 \to V_2$, $r_2 : V_2 \setminus Z_2 \to V_1$ such that the coordinates of $r_i(x,y)$ are rational functions of x and y and $r_1 r_2$, $r_2 r_1$ are the identity on their domains of definition. Such a map r_1 is called[86] a <u>birational map</u> from V_1 to V_2.

It is clear that a birational equivalence between irreducible curves induces an isomorphism between their fields of rational functions, so we have the following.

<u>Theorem 6.6.2</u> *Two irreducible curves are birationally equivalent* (over k) *if and only if their function fields are isomorphic* (over k).

<u>Example 6.6.3</u> We have already met birational equivalence in Example 3.5.6. The field K defined therein is the function field of each of the birationally equivalent curves V, V_1, and V_2 defined by the polynomials $Y^2 - 3X^4 - 1$, $Y_1^2 - X_1^4 - 3$ and $Y_2^2 - 5X_2^3 - 2X_2^2 - 3X_2 - 5$. The equations $X_1 = 1/X$, $Y_1 = 1/X^2$ define a birational map from V to V_1 which is defined except at the points $(0,1)$ and $(0,-1)$.

6.7   PROJECTIVE CURVES

In order to complete the relation between the field K and the geometry, we must move into projective space and add in the 'points at infinity'.

For any field F, let $P^n(F)$ denote n-dimensional projective space over

F, so that $P^n(F) = F^{n+1}\setminus\{0\}/F^*$. The element of $P^n(F)$ represented by the element $(x_0,\ldots,x_n)$ of F^{n+1} is denoted by $(x_0 : \ldots : x_n)$. For each $i = 0,\ldots,n$ the function from $A^n(F)$ to $P^n(F)$ sending $(x_1,\ldots,x_n)$ to $(x_1 : \ldots : x_{i-1} : 1 : x_{i+1} : \ldots : x_n)$ is injective, with image (called an <u>affine piece</u>) $A_i = \{(x_0 : \ldots x_n) \in P^n(F) : x_i \neq 0\}$. We shall often identify $A^n(F)$ with its image.

Now suppose S is a discrete spot on F, with valuation ring R and units U. The spot S defines a topology on F and on F^n for each n. The 'unit ball' R^n in F^n has boundary ∂R^n (the unit sphere) consisting of elements with at least one coordinate in U. The multiplication map $R \times \partial R^n \to F^n\setminus\{0\}$ is continuous and induces a homeomorphism $\partial R^n/U \to P^{n-1}(F)$.

On the other hand, reduction $\bmod\, S$ maps ∂R^n to $F(S)^n\setminus\{0\}$ and so induces a function from $P^{n-1}(F) = \partial R^n/U$ to $P^{n-1}(F(S))$, which is called <u>specialisation</u> at S.

For any homogeneous polynomial $h(Z,X,Y)$ over the perfect field k, let $V(h) = \{(x_0 : x_1 : x_2) \in P^2(k^a) : h(x_0,x_1,x_2) = 0\}$. If $h \neq 0$ $V = V(h)$ is infinite, and is called a <u>projective curve</u>. The intersection V_i of V with A_i is a plane curve consisting of almost all points of V. These plane curves V_0, V_1, V_2 are called <u>affine pieces</u> of V. For example, the curve V_0 is $V(h(1,X,Y)) \subset A^2(k^a)$. If V_0 is irreducible, then so are V_1 and V_2 and we say that V is <u>irreducible</u>. A point P of V is <u>singular</u> if the Jacobian $(\partial f/\partial X_0,\ \partial f/\partial X_1,\ \partial f/\partial X_2)$ vanishes at P, otherwise it is <u>simple</u>. If all points of V are simple, we say V is <u>smooth</u>. Clearly V is smooth if and only if each affine piece V_i is smooth. Moreover if V is smooth, then the polynomial $h(1,X,Y) \in k[X,Y]$ defining V_0 is separable in at least one variable.

If $g \in k[X,Y]$ defines a plane curve $V(g) \subset A^2(k^a)$, and g^H denotes the homogenised polynomial $g^H(X_0,X_1,X_2) = X_0^n g(X_1/X_0,X_2/X_0)$, where n is the total degree of g, then the projective curve $V(g^H)$ meets U_0 in precisely $V(g)$. The complement $V(g^H)\setminus V(g)$ is finite, and consists of the points at infinity. If $V(g)$ is smooth, it does not follow that $V(g^H)$ is smooth.

<u>Example 6.7.1</u> If $g(X) \in k[X]$ is a separable polynomial with distinct roots and $\operatorname{char} k \neq 2$, then the plane curve determined by $f = Y^2 - g(X)$ is smooth, and $V(f^H)\setminus V(f)$ is the single point $(0:0:1)$. This point is simple if and only if $\deg(g) \leq 3$.

Suppose $h(X_0,X_1,X_2)$ is a homogeneous polynomial over k of degree r, such that $V = V(h)$ is an irreducible curve. Since the plane curves V_0, V_1 and V_2 intersect each other in all but a finite subset of each, they are all birationally equivalent. More precisely, the k-algebra homomorphism $k[X_1,X_2] \to k(X_0)[X_2]$ sending X_1 to X_0^{-1} and X_2 to $X_2 X_0^{-1}$ maps the element $h(1,X_1,X_2)$ to a multiple of $h(X_0,1,X_2)$ so induces a homomorphism from $k[V_0] = k[X_1,X_2]/\langle h(1,X_1,X_2)\rangle$ to $k(V_1) = k(X_0)[X_2]/\langle h(X_0,1,X_2)\rangle$, and hence an isomorphism from $k(V_0)$ to $k(V_1)$. The field $k(V_0)$ is called the field of <u>rational functions</u> on V and is denoted by $k(V)$.

Now let $f \in k[X,Y]$ be an irreducible polynomial, separable in Y defining the plane curve $V(f) \subset A^2(k^a)$ and the projective curve $V(f^H) \subset P^2(k^a)$. Let $K = k(V(f))$. We shall define a function $c : P(K) \to V(f^H)/G$ extending $c : P(f) \to V(f)/G$ of §6.2.

Consider the point $(1:X:Y) \in P^2(K)$. If for any $S \in P(K)$ we apply an embedding of $K(S)$ into k^a to the specialisation of $(1:X:Y)$ at S, (which lies in $P^2(K(S))$), then we obtain a point P of $P^2(k^a)$ which lies on $V(f^H)$. We say S <u>lies over</u> P. The point P is well-defined up to the action of G, that is to say, we have a function $c : P(K) \to V(f^H)/G$. If $S \in P(f)$, then $(1,X,Y)$ lies in ∂R_S^3 so c does extend our previous definition. The following theorem is a generalisation of 6.2.2 and 6.3.2.

Theorem 6.7.2   *Let $h(X_0, X_1, X_2)$ be a homogeneous polynomial of degree r
over k defining the irreducible projective curve $V = V(h)$, and let
$K = k(V)$ be its function field.*

*The map $P(K) \overset{c}{\to} V/G$ is surjective. For each point $P \in V$ there are
at most r spots lying over P, and in almost all cases a single spot, the
exceptions being possibly when P is a singular point. In particular, if
V is smooth, then c is a bijection.*

Proof.   This result follows immediately from the observation that under the
isomorphism from $K = k(V_0)$ to $k(V_1)$ described above, the point
$(1, X_1, X_2) \in A^3(K)$ is carried to $(1, X_0^{-1}, X_2 X_0^{-1})$ so that the point
$(1 : X_1 : X_2)$ of $P^2(K)$ is mapped to $(X_0 : 1 : X_2)$ of $P^2(k(V_1))$. For by
6.2.2 we know that for any point $P \in V_1$ there is a spot S such that the
specialisation of $(X_0 : 1 : X_2)$ at S is the orbit of P.

If k is a finite field of order q, then $P^n(k^a)$ is the union of the
subsets $P^n(k_m)$ for $m \in N$, so for any point $P \in P^n(k^a)$ there is a least
m such that $P \in P^n(k_m)$, namely k_m is the subfield generated by the
ratios of any representation of P. We write $k(P)$ for k_m, and put
$\deg(P) = m$. The next theorem follows from 6.7.2 as did 6.5.1 from 6.3.1
and 6.3.2.

Theorem 6.7.3   *Let $h(X_0, X_1, X_2)$ be a homogeneous polynomial over the finite
field k, defining a smooth projective curve $V = V(h) \subset P^2(k^a)$. Let
$K = k(V)$ be the function field of V, and for each $m \in N$, let ν_m be the
number of points in $\{P \in V(h) : k(P) \subset k_m\}$. Then we have*

$$Z(t, K) = \exp \Sigma \, \nu_m \, t^m / m.$$

Remark 6.7.4   The point $P \in V$ has $k(P) \subset k_m$ if and only if P is a fixed point of $\pi^m : V \to V$.   The previous theorem thus relates the Z function of K to the fixed points of the Frobenius map and its iterates on V.

6.8   REMOVAL OF SINGULARITIES

If K is a field with $\mathrm{tr.deg}(K/k) = 1$, then any set $\{X_0, \ldots, X_n\}$ of elements of K which generate it algebraically over k is called a <u>model</u> for K.   To each $S \in P(K)$ and embedding $i : K(S) \to k^a$ we can associate the point $c(S,i) \in P^n(k^a)$ which is the specialisation of $X = (X_0 : \ldots : X_n) \in P^n(K)$ at S.   If h is a homogeneous polynomial over k which vanishes at X, then h vanishes at $c(S,i)$.   If I is the ideal in the ring of homogeneous polynomials over k of functions which vanish at X, then the set

$$V = \{c(S,i) \in P^n(k^a) : S \in P(K) \text{ and } i : K(S) \to k^a \text{ is an embedding}\}$$

of points obtained by this construction is called a <u>projective curve</u>, and is precisely the set of common zeros of all the polynomials in I.   The curve V is also called a model for K.   By Hilbert's Basis Theorem[87], the ideal I is finitely generated, so V is the set of common zeros of a finite set of polynomials.

   In this generalised situation, there are analogous definitions of 'simple point' and 'smoothness' for V in terms of Jacobians of functions in I (see Chapter 9) and as before almost all points of V are simple.   The following theorem[88] indicates the motive behind such a generalisation.

Theorem 6.8.1   *Any field of transcendence degree* 1 *over* k *has a smooth model.*

This theorem and the associated theory of resolution of singularities is beyond the scope of these notes.   However, we give a sketch of the techniques.   If $\{X_0,\ldots,X_n\}$ is an irreducible model[90] for K defining a curve $V \subset P^n(k^a)$ and $S \in P(K)$ lies over a singular point of V, then one chooses by <u>dilation</u> (or <u>blowing up</u>) extra elements $X_{n+1},\ldots,X_m$ (where in fact $m = \frac{1}{2}(n+1)(n+2) - 1$) such that the new model V' corresponding to $\{X_0,\ldots,X_m\}$ is still smooth at the points at spots at which V is smooth, but is now also smooth at S.   Since the set of spots lying over singular points is finite, this process must stop eventually, and one obtains a smooth model.   The generalisation of 6.7.2 states that the smooth model is $P(K)$ (as a point set).

Suppose then that $V \subset P^n(k^a)$ is a smooth model for K.   For each affine piece A_i of $P^n(k^a)$, the intersection $V_i = V \cap A_i$ can be identified with the set of common zeros of polynomials belonging to a certain finitely generated ideal I_i of $R_i = k[X_0/X_i,\ldots,X_n/X_i]$.   The ideal I_i is prime, and the obvious map $R_i \to K$ induces an injective map from $k[V_i] = R_i/I_i$ into K, which exhibits K as the field of fractions of $k[V_i]$.   Thus along with the curve $P(K)$ we have a natural way of identifying elements of K as fractions of polynomial functions defined in each affine piece, i.e. as rational functions on $P(K)$.   In order to make this more precise, we need the language of sheaves, which we delay until Chapter 9.

7 Riemann surfaces

7.1 PUISCEUX'S THEOREM

Let k be an algebraically closed field. If K is a field with tr.deg$(K/k) = 1$, then for each $S \in P(K)$ the field K_S is a finite extension of $k((X))$, so we shall be interested in the algebraic closure of $k((X))$.

For any $n \in N$, the polynomial $Y^n - X$ is irreducible over $k((X))$ by Eisenstein's criterion; let E_n denote its splitting field. The extension $E_n/k((X))$ is separable if and only if $(\text{char } k, n) = 1$, in which case the roots are $\{\omega X^{1/n} : \omega \in \mu_n\}$ where $X^{1/n}$ denotes one of the roots, the degree of the extension is n, and the Galois group is cyclic and can be identified with μ_n in the obvious way. Since E_n is (isomorphic to) the field of Laurent series over k in any one of the roots, we write $k((X^{1/n}))$ for E_n.

Proposition 7.1.1 *If* k *is an algebraically closed field and* $E/k((X))$ *is an extension of degree* n *with* $(\text{char } k, n) = 1$, *then* $E/k((X))$ *is isomorphic to* $k((X^{1/n}))/k((X))$.

Proof Since k is algebraically closed, the relative degree of this extension in 1, so the ramification index is n by 4.6.6. If U is the group of units at the spot on E and $t \in E$ is a prime element, then $Xt^{-n} \in U$, but any element of U has n distinct n-th roots by Hensel's lemma (4.5.5), so $Y^n - X$ splits completely over E.

Definition 7.1.2 The field $k|X| = \varinjlim k((X^{1/n}))$ is called the field of fractional Laurent series in X over k. For $h \in k|X|$, the least $n \in N$

such that $h \in k((X^{1/n}))$ is called the <u>uniformity</u> of h.

As a simple corollary of 7.1.1 we have Puisceux's Theorem[91].

<u>Theorem 7.1.3</u>   *If* k *is an algebraically closed field of characteristic* 0, *then* $k|X|$ *is the algebraic closure of* $k((X))$.

In this case, there is by 7.1.1 for each $n \in N$ a unique subfield of $k|X|$ of degree n over $k((X))$, namely $k((X^{1/n}))$, so the Galois group of $k|X|/k((X))$ is the profinite group $\hat{Z} = \varprojlim \mu_n$ (cf. §6.5 for finite fields). It is easy to see that if F is a field of characteristic 0 but not necessarily algebraically closed, then the algebraic closure of $F((X))$ is the subfield of $F^a|X|$ consisting of those fractional Laurent series whose coefficients generate a finite extension of F.   The following trivial corollary (of 7.1.1 and the above remarks about the Galois group) will be needed.

<u>Corollary 7.1.3</u>   *If* k *is an algebraically closed field of characteristic* 0 *and* $f(Y)$ *is an irreducible monic polynomial of degree* n *over* $k((X))$ *then the splitting field of* f *is* $k((X^{1/n}))$.   *The roots of* f *have uniformity* n, *and there is an element* $h(t) \in k((t))$ *such that*

$$f(Y) = \prod_{\omega \in \mu_n} (Y - h(\omega X^{1/n})).$$

By abuse of notation, we refer to the element $h \in k((t))$ of this corollary as a <u>root</u> of f.

Characteristic 0 is crucial to Puisceux's theorem.   For if k is an algebraically closed field of characteristic $p \neq 0$, the separable extensions of $k((X))$ in $k|X|$ have degree prime to p although $k((X))$ has separable extensions of degree p, for example[92] the splitting field of $Y^p - Y - X^{-1}$.   Puisceux's Theorem allows for the replacement in character-istic 0 of the theory of spots and valuations by the theory of fractional

Laurent series, as we indicate in the next section.

7.2   PARAMETRISATIONS AND SPOTS

In this section, k will be an algebraically closed field of characteristic
0.   Let $f(X,Y) \in k[X,Y]$ be an irreducible polynomial defining the curve
$V = V(f) \subset A^2(k)$, and let $K = k(V)$ be the function field of V.   Each
spot $S \in P(K)$ determines a point on V by 6.2.2.   Suppose $S \in P(K)$ lies
over $(x,y) \in V$, and let $g \in k[[X-x]][Y]$ be the corresponding factor of
f over $k[[X-x]]$.   We assume without loss of generality that g is monic.
The degree m of g is equal to the ramification index of the spot S, and
to the uniformity of each root of g by 7.1.3.   Let $h \in k((t))$ be a root
of g, so that $g(t^m + x,Y) = \prod\limits_{\omega \in \mu_m} (Y - h(\omega t))$.   Since the roots of g are
integral (see the proof of 6.2.2) h belongs to $k[[t]]$.   Substituting
$t = 0$, $Y = y$ we see that $h(0) = y$.

<u>Definition 7.2.1</u>   A <u>parametrisation</u> of the curve V defined by $f(X,Y) = 0$
near the point (x,y) is a pair (a,b) or non-constant elements of $k[[t]]$
such that $f(a,b) = 0$ (in $k[[t]]$) and $a(0) = x$, $b(0) = y$ in k.

We have seen above that each spot $S \in P(K)$ determines a parametrisation
near the point over which it lies.   Conversely if (a,b) is a parametrisation of V near (x,y) then the elements $a,b \in k((t))$ are transcendental over k, so there is an embedding of $K = k(X)[Y]/\langle f \rangle$ into $k((t))$
sending X to a, Y to b, which defines a spot on K.   There is an
obvious equivalence relation between the parametrisations near a point
corresponding to change of variable; an equivalence class is called a <u>place</u>
or <u>branch</u> at (x,y), the point (x,y) then being called the <u>centre</u> of the
place.   The correspondence sketched above between spots and places is
bijective.

More generally, suppose K is a field of transcendence degree 1 over k and $\{X_0, \ldots, X_n\}$ is a model for K defining a curve $V \subset P^n(k)$. If S is a spot in $P(K)$ lying over the point $x = (x_0 : \ldots : x_n) \in P^n(k)$ of V let $r \in K$ be such that $(rX_0, \ldots, rX_n) \in \partial R_S^{n+1}$. Now K_S is (isomorphic to) the field $k((t))$ where $t \in K$ is a prime element (see (4.5.4)); let $h_i \in k[[t]]$ be the element rX_i. Since $h_i(0) = x_i$, the sequence $(h_0, \ldots, h_n)$ of elements of $k[[t]]$ is a <u>parametrisation</u> of V near P in the obvious generalised sense.

7.3   CURVES AS RIEMANN SURFACES

We now interpret some of the results of Chapter 6 and sections 7.1, 7.2 in the case $k = C$, which leads to the familiar classical theory of Riemann surfaces and complex curves. We do this to provide a concrete example of the abstract theory, which provides a geometric insight to the results and techniques valid for arbitrary fields (particularly finite fields) and at the same time provides an introduction to the theory of complex manifolds which we shall also need. The essential ingredient is the introduction of topology and complex analysis by virtue of the usual topology on C, which is induced by the infinite complex affine and projective spaces, and their subsets. In particular, any curve carries a complex topology, and, being the set of zeros of a finite family of polynomials, is closed in its ambient space. In particular any projective curve is compact since projective spaces are compact. It is a non-trivial result that irreducible curves are connected, and we delay the proof of this until section 7.5.

We recall that if a power series $h \in C[[t]]$ has a positive radius of convergence R, then the function $z \mapsto h(z)$ from the disc $\{z \in C : |z| < R\}$ to C is holomorphic. The subring of elements of $C[[t]]$ with positive radius of convergence is denoted by $C_0[[t]]$, its elements can be identified with germs of holomorphic functions at the origin. Its field of fractions, denoted by $C_0((t))$, is the field of germs of meromorphic functions at the origin. The t-adic spot on $C((t))$ restricts to a spot on $C_0((t))$ whose maximal ideal is generated by t and consists of germs of holomorphic functions which vanish at the origin. The following theorem provides the link between algebra and analysis[93].

<u>Theorem 7.3.1</u>   *The field $C_0((t))$ is dense and algebraically closed in $C((t))$.*

If we put $C_0|t| = \lim_{\rightarrow} C_0((t^{1/n}))$, then by Puisceux's theorem (7.1.2) and 7.3.1 we see that $C_0|t|$ is the algebraic closure of $C_0((t))$. Let $f \in C[X,Y]$ be an irreducible polynomial defining the curve $V \subset C^2$ with function field K. Suppose $S \in P(K)$ lies over the point $(x,y) \in V$ and has ramification index m over the spot x on $k(X)$. Let g be the corresponding factor of f over $C[[X-x]]$, and let h be a root of g. By 7.3.1 h belongs to $C_0[[t]]$ so that $z \longmapsto (z^m+x,h(z))$ is a holomorphic map from a sufficiently small neighbourhood of the origin in C to V taking 0 to (x,y). The Jacobian of this map is $(mz^{m-1},dh/dz)$, which is non-zero away from the origin. If the spot S is unramified so that $m = 1$, the Jacobian is also non-zero at the origin.

<u>Proposition 7.3.2</u>   *Suppose (x,y) is a simple point of V and $S \in P(K)$ lies over (x,y) and has ramification index m over $C(X)$ with $m \geq 2$. If $h \in C[[t]]$ is a root of S, then dh/dt is non-zero at the origin.*

<u>Proof</u>   If g is the irreducible factor of f over $C[[X-x]]$ corresponding to S, then we can write $f = ga$ where $g,a \in C[[X-x]][Y]$, so that
$$f(t^m+x,Y) = b \prod_{\omega \in \mu_m} (Y - h(\omega t)) \quad \text{for some } b \in C[[t]][Y].$$
Suppose ord $h = r$. Differentiating this equation and putting $Y = y$ we see that

$\text{ord}(\partial f/\partial t) \geq mr - 1$, and $\text{ord}(\partial f/\partial Y) \geq m - 1$. Since $\partial f/\partial t = mt^{m-1}\partial f/\partial X$, we see that $\text{ord}(\partial f/\partial X) \geq mr - m$. So since $m \geq 2$, $\partial f/\partial Y$ vanishes at (x,y), and if $r \geq 1$ (which is precisely when dh/dt vanishes at 0) $\partial f/\partial X$ also vanishes at (x,y).

So the map $z \mapsto (z^m + x, h(z))$ is a local isomorphism at the origin if $m = 1$ or if the point (x,y) is simple, but in all cases it is a local isomorphism away from the origin. Thus in a neighbourhood of a simple point, the curve has a holomorphic parametrisation, so if the curve is smooth it carries the structure of a complex manifolds. Similar arguments apply to curves in the projective plane, or more generally in projective space, and so for any field K with $\text{tr.deg}(K/C) = 1$, we can consider $X = P(K) \subset P^n(C)$ as a compact Riemann surface as indicated in §6.8. For each affine piece U_i of $P^n(C)$, the intersection $X_i = X \cap U_i$ is a curve in C^n and the elements of K as rational functions on X_i are meromorphic. That is to say, we can consider elements of K as meromorphic functions on $P(K)$. For example, if $K = C(t)$ then as a set $P(K) = P^1(C)$ by 4.1.11, and the Riemann surface structure on $P^1(C)$ is the usual one.

The construction of the Riemann surface $P(K)$ is natural, so any extension L/K determines a holomorphic map between the Riemann surfaces, we consider the topology of a curve near a singular point.

For simplicity, suppose V is a curve in C^2 defined by the polynomial $f \in C[X,Y]$ with $(0,0)$ as a singular point. Factorising f over $C[[X]]$ we have $f = g \prod_i f_i$ where the irreducible factors f_i vanish at $(0,0)$ and correspond to the spots S lying over $(0,0)$ and where $g(0,0) \neq 0$. Let $h_i \in C[[t]]$ be a root of f_i, and let m_i be the degree of f_i. Choose disc neighbourhoods U, U' of the origin in C such that

i) h_i is holomorphic in the disc $\{z \in C: z^{m_i} \in U\}$,

ii) $(0,0)$ is the only singular point of $V \cap (U \times U')$,

iii) g is holomorphic and does not vanish in $U \times U'$.

In particular, g and the factors f_i are all holomorphic in $U \times U'$. Let (x,y) be a simple point of $V \cap (U \times U')$. Substituting $X = x$, $Y = y$ in the equation $f = g \prod f_i$ (between continuous functions in $U \times U'$), we see that $f_i(x,y) = 0$ for some i (since $g(x,y) \neq 0$) and since (x,y) is a simple point such a value of i is uniquely determined. If as before we write g for f_i, h for h_i and m for m_i, then $g(t^m,Y) = \prod_{\omega \in \mu_m} (Y - h(\omega t))$, so choosing t such that $t^m = x$ we see that $y = h(\omega t)$ for some (unique) $\omega \in \mu_m$. Putting $z = \omega t$ we have $(x,y) = (z^m,h(z))$.

<u>Lemma 7.3.3</u> *The map* $z \mapsto (z^m,h(z))$ *is a local homeomorphism at the origin.*

<u>Proof</u> By 7.3.2 we can assume without loss of generality that $m > 1$. If this map is not a local homeomorphism, then there are two sequences $\{a_n\},\{b_n\}$ of points tending to 0 with $a_n \neq b_n$ and $(a_n^m,h(a_n)) = (b_n^m,h(b_n))$. So for each n there is an element $\omega_n \in \mu_m \setminus \{1\}$ such that $a_n = \omega_n b_n$. Hence there is an element $\omega \in \mu_m$ with $\omega \neq 1$ and a subsequence $\{c_n\}$ of $\{a_n\}$ such that $h(c_n) = h(\omega c_n)$ for all n. Since h is holomorphic this implies that $h(t) = h(\omega t)$ near the origin which contradicts the fact that h has uniformity m.

Of course a holomorphic map can be a local homeomorphism without being a local isomorphism (e.g. $z \mapsto (z^2,z^3)$). Putting the above observations together we have the following.

<u>Proposition 7.3.4</u> *Suppose* $f \in C[X,Y]$ *is an irreducible polynomial defining the plance curve* $V \subset C^2$. *If* $P = (0,0)$ *is a singular point of* V *and* $S_1,\dots,S_r$ *are the spots lying over* P, *with* m_i,h_i *the ramification index*

and root of S_i, *then there is a neighbourhood* U *of* O *in* C *such that*

  i) *the function* h_i *are all holomorphic in* U,

  ii) *if* W_i *denotes the image of* U *under the map* $z \mapsto (z^{m_i}, h_i(z))$ *then* $W_i \cap W_j = \{P\}$ *and* $W = \cup W_i$ *is a neighbourhood of* P *in* V,

  iii) *the map* $\coprod_i h_i : \coprod_i U \to V$ *induces a homeomorphism* $\overset{r}{\underset{i=1}{V}} U \to W$, *where* $\overset{r}{\underset{i=1}{V}} U$ *denotes the union of* r *copies of* U *with their origins identified.*

Remark 7.3.4   A curve V with a singular point P cannot be[94] a smooth manifold in any neighbourhood of P.

7.4   HOLOMORPHIC THEORY OF RIEMANN SURFACES

We briefly recall some of the basic results about Riemann surfaces.   In order to abbreviate the language, we resort to category notation.   Let RS be the category of Riemann surfaces (or complex manifolds of dimension 1) and non-constant holomorphic maps.   We use the word 'isomorphism' for invertible maps in this category.   Let RS_0 be the full subcategory of compact connected Riemann surfaces.   The Riemann sphere $S^2 = P^1(C)$, which as usual we consider as the one-point compactification $C \cup \infty$ of C with $\infty = (1:0) \in P^1(C)$, belongs to RS_0.   As usual, by a **meromorphic function** on the Riemann surface X is we mean any holomorphic map $F: X \to S^2$ apart from the constant map to ∞; the elements of $f^{-1}(\infty)$ are then called the **poles** of f.   The pointwise sum and product of meromorphic functions is again meromorphic.   If $X \in Ob(RS)$ and $x \in X$ we say that u is a **local parameter centred at** x if u is an isomorphism from some neighbourhood U of O in C to its image in X with $u(O) = x$.

    Suppose $f: X \to Y$ is a map in RS, and $x \in X$.   Let u', u be local parameters centred at $x, f(x)$.   The composite $h = u^{-1} f u'$ is a holomorphic

94

function in some neighbourhood of 0 in C with $h(0) = 0$. The order $n(f,x)$ of h at 0 is a non-negative integer independent of the choice of u,u' (see §7.6) and is called the <u>analytic ramification index</u> of f at x. If $n(f,x) = 1$ we say f is <u>unramified</u> at x, otherwise f is <u>ramified</u> at x. Clearly f is a local isomorphism at x if and only if f is unramified at x. Now we can write $h(z) = az^n(1 + g(z))$ where $n = n(f,x)$, $a \in C$ is non-zero and $g(0) = 0$. In a suitably small neighbourhood of 0 there is a holomorphic function $b(z)$ such that

$$b(z)^n = a(1 + g(z)), \quad \text{namely} \quad b(z) = \sqrt[n]{a} \sum_r \binom{1/n}{r} g(z)^r, \quad \text{where} \quad \sqrt[n]{a} \text{ is any}$$

n-th root of a. Since the map $i: z \mapsto zb(z)$ is a local isomorphism at 0, we can take $v = ui^{-1}$ as a local parameter in Y centred at $f(x)$, and in this case, $v^{-1}fu'(z) = z^n$. This describes the local behaviour of f at x. We note that the integer n is the topological degree of f at x. The following proposition is now trivial.

<u>Proposition 7.4.1</u> *If $f: X \to Y$ is a non-constant holomorphic map between Riemann surfaces, then f is open. For each point $x \in X$ there is a neighbourhood U of x in X such that $f: U \to f(U)$ is a local isomorphism away from x and for each point $y \in f(u) \setminus \{f(x)\}$, the set $U \cap f^{-1}(y)$ has order $n(f,x)$. The set of points at which f is ramified is discrete. If X,Y are compact and connected then f is surjective.*

Suppose $f: X \to Y$ is a map in $\underline{RS}_0$. By 7.4.1 we know that for each $y \in Y$ the set $f^{-1}(y)$ is non-empty, finite (otherwise there would be a limit point $x \in f^{-1}(y)$ and no neighbourhood U of x could satisfy the conditions of 7.4.1), and the integer $n = \sum_{x \in f^{-1}(y)} n(f,x)$ is independent of y. This integer is called the <u>valence</u> of f and is the number of times, with multiplicities, that f takes each value in Y. If $D \subset X$ is the set of ramification points of f, then D is finite (by 7.4.1) and $f: X\backslash D \to Y\backslash f(D)$ is an n-fold covering space. In particular, if X is in $\underline{RS}_0$ and $f: X \to S^2$ is a meromorphic function, then the valence of f is the number of zeros (or poles) counting multiplicities.

For $X \in Ob(\underline{RS})$ and $U \subset X$ open, let $H_X(U)$, $M_X(U)$ denote the rings of holomorphic and meromorphic functions on U. For $x \in X$, the limits $H_{X,x} = \lim_{\to} \{H_X(U): U \text{ open}, \ x \in U\}$ and $M_{X,x} = \lim_{\to} \{M_X(U): U \text{ open}, x \in X\}$ are the rings of germs of holomorphic and meromorphic functions at x. A map $f: X \to Y$ in $\underline{RS}$ induces by composition ring homomorphisms $f_x: H_{Y,f(x)} \to H_{X,x}$ and $f_x: M_{Y,f(x)} \to M_{X,x}$. In particular, a local parameter $u: U \to X$ centred at x induces an isomorphism between the pairs $H_{X,x} \subset M_{X,x}$ and $C_0[[t]] \subset C_0((t))$. The ring $M_{X,x}$ is the field of fractions of the local domain $H_{X,x}$ and so carries a natural spot $S(x)$, namely the m_x-adic spot, where m_x is the unique maximal ideal of $H_{X,x}$ (the kernel of the evaluation map $H_{X,x} \to C$, $f \mapsto f(x)$).

Now suppose $X \in Ob(\underline{RS}_0)$. By Liouville's theorem , $H_X(X) = C$. The ring $M_X(X) = M(X)$ is a field. It is well-known[95] that $M(S^2)$ is the field $C(t)$ of rational functions in t, where $t: S^2 \to S^2$ is the identity map. For $x \in X$, the restriction map $M(X) \to M_{X,x}$ associates to a meromorphic function its Laurent expansion at x, and induces a spot $S(x)$ on $M(X)$ whose valuation ring consists of those meromorphic functions on X which do not have a pole at x. The following non-trivial result[96] is crucial.

<u>Theorem 7.4.2</u>   *The non-constant meromorphic functions on a compact Riemann surface* X *separate points.   In addition, for each* $x \in X$ *there is a holomorphic map* $f: X \to S^2$ *with* $f(x) = 0$ *which is a local isomorphism at* x.

96

It follows from 7.4.2 that if x_1, x_2 are distinct points of X, then $S(x_1) \neq S(x_2)$, so that S is an injective function from X to the set $P(X)$ of non-trivial spots on $M(X)$ which are trivial on C. We shall see presently (7.5.3) that S is in fact bijective.

<u>Lemma 7.4.3</u> *The restriction map* $M(X) \to M_{X,x}$ *becomes an isomorphism on completion at the spot* $S(x)$.

<u>Proof.</u> By 7.4.2 it suffices to prove this result for $X = S^2$, when it is trivially true since $C(t) \subset C_0((t))$ becomes an isomorphism on completion.

We are now ready to prove an essential result. A map $f: X \to Y$ in $\underline{RS}_0$ induces by composition a homomorphism $f^*: M(Y) \to M(X)$.

<u>Proposition 7.4.4</u> *If* $f: X \to Y$ *is a non-constant holomorphic map between compact connected Riemann surfaces of valence* n, *then each element* $g \in M(X)$ *is a root of a monic polynomial over* $M(Y)$ *of degree n, whose coefficients have no poles outside* $f(g^{-1}(\infty))$.

<u>Proof.</u> Let $D = \{x \in X: f$ is ramified at x or $g(x) = \infty\}$ so that $f: X \backslash D \to Y \backslash f(D)$ is n-fold covering space. For $y \in Y \backslash f(D)$, let $s_r(y)$ be the r-th elementary symmetric function in the n complex numbers $\{g(x): x \in f^{-1}(y)\}$. If U is an open set containing y such that $f^{-1}(U)$ is the disjoint union of n open sets $U_1, \ldots, U_n$ each mapping isomorphically under f onto U, and $h_i: U \to U_i$ is the inverse to f, then (restricted to U) s_r is the r-th elementary symmetric function in the n holomorphic functions $gh_1, \ldots, gh_n$. Hence s_r is a holomorphic function in $Y \backslash f(D)$.

Suppose $y \in f(D)$ is not the image of a pole of g. Each point $x \in f^{-1}(y)$ has a neighbourhood containing no poles of g and so g is bounded near $f^{-1}(y)$, and hence s_r is bounded near y. By Cauchy's theorem[97], y is a removable singularity of s_r for each r.

If $y \in f(D)$ is the image of a pole of g, the function s_r may be

bounded near y, but at worst it may have a pole at y.   In any case we see
that each s_r is a meromorphic function on Y.

The complex number g(x) is a root of the polynomial
$$Z^n = s_1 f(x) Z^{n-1} + \ldots + (-1)^n s_n f(x)$$ for each $x \in X \setminus g^{-1}(\infty)$ and so g is a
a root of the polynomial $Z^n - s_1 f Z^{n-1} + \ldots + (-1)^n s_n f$ since[98] two
holomorphic maps which agree almost everywhere are equal.

<u>Corollary 7.4.5</u>   *If* f: X → Y *is a non-constant holomorphic map between
compact connected Riemann surfaces of valence* n,   *then* M(Y)/M(X) *is a
finite extension of degree* n.

<u>Proof</u>   By standard field theory $\deg(M(X)/M(Y)) \leq n$ since every element of
M(X) is algebraic over M(Y) of degree $\leq n$.   Let y be a point of Y
such that $f^{-1}y$ has order n.   Since the functions in the algebra M(X)
separate points (by 7.4.2) there is a meromorphic function on X which takes
n distinct values on the set $f^{-1}y$, and such a function cannot satisfy a
polynomial equation over M(Y) of degree less than n.

<u>Corollary 7.4.6</u>   *If* X *is a compact connected Riemann surface, then* M(X)
is a field of transcendence degree 1 *over* C.   *If* g,h *are meromorphic
functions on* X *of valencies* m,n *then there is a polynomial* $F \in C[t,u]$
of degree at most n *in* t *and* m *in* u *such that* F(g,h) = 0 *in* M(X).

<u>Proof</u>   By 7.4.2 there is a non-constant holomorphic map $h: X \to S^2$ so that
M(X) is a finite extension of $M(X^2) = C(t)$,   and hence $\mathrm{tr.deg}(M(X)/C) = 1$.

Now consider the extension $h^*: M(S^2) \to M(X)$ of degree n.   By 7.4.4
there is a polynomial over $M(S^2)$ of degree n with g as a root, and
hence an irreducible polynomial $F \in C[t,u]$ of degree at most n in t with
F(g,h) = 0 in M(X).   Since F is then the minimal polynomial connecting
g and h, we see by symmetry that F has degree at most m in u.

98

We have already seen in 7.3 that if K is a field with $\mathrm{tr.deg}(K/C) = 1$ then the non-singular model $P(K)$ is a compact Riemann surface.   In order to show that $P(K)$ is connected, we need the following simple lemma.

__Lemma 7.5.1__   *If E is a field with valuation v and $e \in E$ is a root of the polynomial $X^n + a_{n-1} X^{n-1} + \ldots + a_0 \in E[X]$, then $v(e) \leq 1 + \max_i v(a_i)$.*

__Proof__   Since $e^n = -\Sigma a_i e^i$, we have $v(e)^n \leq \Sigma v(a_i)(v(e))^i$, so if $r = v(e)$ and $C = \max v(a_i)$, then $r^n \leq C(1+r+\ldots+r^{n-1})$ and elementary algebra shows that $r \leq 1 + C$.

__Theorem 7.5.2__   *If K is a field of transcendence degree 1 over C then the Riemann surface $P(K)$ is connected.*

__Proof__   Let $t \in K$ be an element such that $K/C(t)$ is a finite extension of degree n say, and let $f: X = P(K) \longrightarrow S^2 = P(C(t))$ be the map induced between Riemann surfaces.   Assume for a contradiction that X is not connected, and let Y be a component, so that the valence r of $f: Y \to S^2$ is less than n.   Let u be a primitive element of K over $C(t)$ with $g \in C[t,u]$ the irreducible polynomial such that $g(t,u) = 0$ in K.   If $g_n(t)$ is the leading coefficient, then $v = g_n(t)u$ is a primitive element, and is integral over $C[t]$, so as a meromorphic function on Y, has its poles in $f^{-1}(\infty)$.

If $Z^r + a_{r-1} Z^{r-1} + \ldots + a_r$ is the minimum polynomial of v over $C(t)$, then since v in integral, each a_i is a polynomial.   By 7.5.1 for each $y \in Y \backslash f^{-1}(\infty)$ we have $|u(y)| \leq C|f(y)|^m$ for some $C > 0$ where m is the maximum of the degrees of the polynomials a_i.   Hence $|s_r(z)| \leq C^r|z|^{rm}$ for all points $z \in C$ over which f is unramified, so $|s_r(z)| \leq D|z|^{rm}$ for all $z \in C$ for suitable D.   But s_r is holomorphic in C by 7.4.4, and so is a polynomial[99], and v satisfies a polynomial equation of degree r over $C(t)$, which is a contradiction.

<u>Corollary 7.5.3</u>   *If* X *is a compact connected Riemann surface,* *the natural*
map $S: X \to P(X) = P(M(X))$ *is an isomorphism.*

<u>Proof</u>   We already know from §7.4 that S is injective.   Since P(K) is
connected by 7.5.2 and S is holomorphic, S is surjective and hence an
isomorphism.

<u>Theorem 7.5.4</u>   *The functors* $X \longmapsto M(X)$, $K \longmapsto P(K)$ *are inverse equivalences*
between the category $\underline{RS}_0$ *and the category of fields of transcendence degree*
1 *over* C.

<u>Remark 7.5.5</u>   Each of these categories is also equivalent to the category of
smooth projective curves over C with suitably defined morphisms (whose
definition we delay until Chapter 9).

The proof of 7.5.2 also shows that an irreducible affine curve is also
connected.   However, not every non-compact connected Riemann surfaces is a
curve.   For example, the open unit disc U is not a curve.   Any affine
curve can be obtained by removing a finite set of points from a projective
curve, so if U were a curve, then we would have a finite covering space
$f: U \backslash D \to C \backslash f(D)$ as above.

The elementary symmetric functions in the inclusion map $i: U \hookrightarrow C$ are
holomorphic in C but are bounded, so are constant by Liouville's theorem,
which is a contradiction.

7.6   RAMIFICATION REVISTISTED

We can now reinterpret the analytic ramification index of a map $f: X \to Y$ in
$\underline{RS}_0$ by virtue of the previous section.   Suppose $x \in X$ and $y = f(x)$.   The
spot S(y) on M(Y) extends the spot S(x) on M(X) and the analytic
ramification index $n = n(f,x)$ is precisely the ramification index e of

S(y) over S(x). For if we choose a local parameter u centred at y inducing an isomorphism $M_{Y,y} \to C_0((t))$, then since the extension $M_{X,x}/M_{Y,y}$ has degree e (by 7.4.3 and 4.3.7) it is isomorphic to $C_0((t^{1/e}))/C_0((t))$ (by 7.1.1 and 7.3.1) so there is a local parameter u' centred at x such that $v^{-1}fu(z) = z^e$, and so e = n.

__Proposition 7.6.1__ *If* L/K *is a finite extension of fields of transcendence 1 over* C *corresponding to a map* $f: X \to Y$ *between the Riemann surfaces* X = P(K), Y = P(L), *then* $f: X \to Y$ *is a covering space if and only if* L/K *is unramified.*

If K is a field with $\mathrm{tr.deg}(K/C) = 1$ and $t \in K$ is an element such that $K/C(t)$ is a finite extension of degree n, then the integer

$$\left[\frac{1}{2} \sum_{S \in P(C(t))} \sum_{T \in S^k} (e_{T|S} - 1) \right] + 1 - n$$

is independent of the choice of t (compare 4.7.12) and is called the

__algebraic genus__ of K.   The next result, the Riemann-Hurwitz theorem, is the

first in a series of 'index theorems'[100].

__Theorem 7.6.2__ *If* K *is a field with* $\mathrm{tr.deg}(K/C) = 1$, *then the algebraic genus* g_a *of* K *is equal to the topological genus* g_t *of the Riemann surface* P(K).

__Proof__   Suppose $K/C(t)$ is a finite extension of degree n and that $P(K) \to S^2$ is the corresponding map between Riemann surfaces.   If $D \subset X$ is the set of ramification points of f, then we can choose triangulations of X and S^2 such that

  i) each point of D is a vertex,

 ii) f is simplicial,

iii) no edge joins vertices in f(D),

 iv) for each simplex σ of X, $f: \sigma \to f(\sigma)$ is homeomorphism.

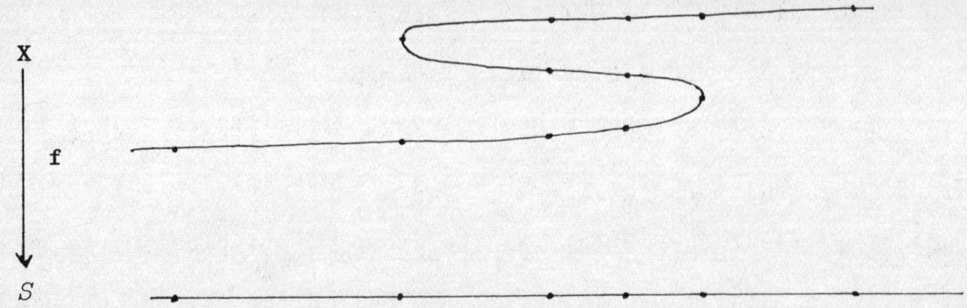

Such a choice is possible since the set D is finite, $f: X \backslash D \to S^2 \backslash f(D)$ is

a covering space and of course since surfaces can be triangulated[101] . If a

triangulation of a compact surface has e_i i-simplices, then the integer

$e_2 - e_1 + e_0$ is equal to[102] the Euler characteristic 2 - 2g. Now since each

simplex of S^2 except the vertices in f(D) is covered by precisely n

simplices in X of the same dimension, and S^2 has topological genus 0,

we see that

$$2 - 2g_t = n(2 - 0) - \sum_{x \in D} n(f,x)$$

$$= 2n - \left[\sum_{s(y) \in P(M(S^2))} \sum_{x \in f^{-1}(y)} (e_{S(x)|S(y)} - 1) \right]$$

and the result follows.

7.7   DIVISORS

Let $X \in \mathrm{Ob}(RS_0)$. A <u>divisor</u> on X is an element of Div(X), the free

abelian group on the points of X, or equivalently (by 7.5.3) on the spots

in P(X). The augmentation $\sum n_x x \mapsto \sum n_x$ from Div(X) to $\mathbb{Z}$ is called the

<u>degree map</u> (cf. §5.4) and its kernel is denoted by $Div_0(X)$.

For any meromorphic function $f: X \to S^2$ on X, the sets of zeros and poles are finite. A point $x \in X$ is a zero (pole) of f if and only if

$\text{ord}_x(f) > 0$ (< 0), where $\text{ord}_x: M(X) \to Z$ associated to each $f \in M(X)$ its

order at x and determines (see (4.5.1)) the spot $S(x) \in P(X)$. The

divisor $(f) = \Sigma \text{ord}_x(f)x$ (zeros minus poles of f) lies in $\text{Div}_0(X)$ (see

§7.4). The function $f \mapsto (f)$ from $M(X)^*$ to $\text{Div}_0(X)$ is a homomorphism

with kernel C^*; its image $\text{Pr}(X)$ is the group of <u>principal divisors</u> on X.

The quotient group $J(X) = \text{Div}_0(X)/\text{Pr}(X)$ is called the <u>divisor class group</u>,

or <u>Jacobian</u> of X. We shall see in §11.4 that $J(X)$ also has an inter-

pretation in terms of line bundles over X (cf. §5.4).

For any topological space Y, let $Y^{(n)}$ denote the n-fold symmetric

product Y^n/S_n of Y. The elements of $Y^{(n)}$ are unordered n-ples of

elements of Y. Each $x = (x_1,\ldots,x_n) \in X^{(n)}$ determines an element

$i(x) = \Sigma x_j \in \text{Div}(X)$ of degree n, so there is a natural map

$i: X^{(n)} \to \text{Div}(X)$. If we choose a base point $* \in X$, and base $X^{(n)}$ at

$* = (*,\ldots,*)$, then the function $x \mapsto i(x) - i(*)$ induces a function from

$X^{(n)}$ to $J(X)$ taking the base point to 0.

It is a consequence[103] of the Riemann–Roch theorem that if g is the

genus of X, then $X^{(g)} \to J(X)$ is bijective. That is to say, for any

points $x,y \in X^{(g)}$ there is a unique point $z \in X^{(g)}$ such that

$i(x) + i(y) - i(z) - i(*) = (f)$ for some $f \in M(X)$, where $(0) = 0$ by

definition.

Now $\text{Div}(X)$ can be given a unique topology so that it is the free

abelian <u>topological</u> group on the space X. This topology induces a topology

on $J(X)$ making it into a topological group. Of more importance is an

alternative description of this topology. For any $n \in N$, the space $X^{(n)}$

is naturally a complex n-dimensional manifold[104]. The reason for this,

roughly speaking, is that X is locally isomorphic to C, and so $X^{(n)}$ is

103

locally isomorphic to $C^{(n)}$ and because of the following lemma.

<u>Lemma 7.7.1</u>   *If* $s_i : C^n \to C$ *is the* i-*th elementary symmetric function then* $s = (s_1, \ldots, s_n) : C^n \to C^n$ *induces a homeomorphism* $C^{(n)} \to C^n$.

<u>Proof</u>   If $a \in C^n$ has image $s \in C^n$ then we have $\Pi(1 + a_i t) = \Sigma x_i t^i$.   The fact that $s : C^{(n)} \to C^n$ is bijective follows from the fact that C is algebraically closed.   The inverse of s is continuous by 7.5.1.

The map $X^{(g)} \to J(X)$ is in fact a homeomorphism, so $J(X)$ is a compact Lie group; being a finite dimensional and abelian it is a complex torus[105]. The natural map $x \mapsto (x, *, *, \ldots, *)$ from X into $J(X)$ is holomorphic, and is universal with respect to holomorphic maps into complex tori.

We note that $J(X)$ as a set can be defined in terms of the field $M(X)$, independently of the complex topology.   In fact $J(X)$ is a projective variety[106], and the group operations are algebraic.

7.8   APPENDIX: THE FUNDAMENTAL GROUP

We have already seen in §7.4 that if K is a field with $\mathrm{tr.deg}(K/C) = 1$ and $X = P(K)$ then any finite unramified extension L/K of K determines a finite connected covering space $f : Y \to X$ in $\underline{\mathrm{RS}}_0$ with $Y = P(L)$ of degree $\deg(L/K)$.   Conversely, suppose $f : X' \to X$ is a finite connected covering space of X or degree n.   There is a unique Riemann surface structure on X' such that $f : X' \to X$ is a map in $\underline{\mathrm{RS}}_0$, and then $M(X')/M(X)$ is an unramified extension of $K = M(X)$ of degree n.   Thus there is an equivalence between the category of finite unramified extensions of K and the category of finite connected covering spaces of X.   Normal extensions of K correspond to regular covering spaces of X and its addition, if L/K is a normal extension corresponding to the covering space $f : Y \to X$, then the Galois group L/K is isomorphic to the group of covering transformations of

$f: Y \to X$, which itself is isomorphic to the quotient group $\pi_1(x)/f_*\pi_1(Y)$. Now the n-fold connected covering spaces of X are classified[107] by the subgroups of index n in and so we have the following result.

<u>Theorem 7.8.1</u> *Let* K *be a field with* $\mathrm{tr.deg}(K/C) = 1$. *The 'Galois group'* $G = \lim_{\leftarrow} \{\mathrm{Gal}(L/K): L/K$ *a finite normal unramified extension*$\}$ *is naturally isomorphic to the profinite completion* $\pi_1(X)^\wedge$ *of the fundamental group of Riemann surface* $X = P(K)$.

Once again we have a connection with class field theory: if there were a maximal unramified extension E/K in K^a/K, then G would be its Galois group. Such a field E would be the analogue of the universal covering space of X. Except when X has genus 0, the universal covering space of X is non-compact so E does not exist.

<u>Exaple 7.8.2</u> Consider the field $K = C(t)$, so that $X = S^2$. If L/K is a normal unramified extension of degree $n \geq 2$, then L is the splitting field of some polynomial $f \in C[t,u]$ over $C(t)$. The discriminant $D(t)$ of this polynomial in u, is a polynomial in t and over any root z of $D(t)$ in S^2 the extension is ramified (cf. 4.7.9). Hence there are no finite normal unramified extensions, and hence no finite covering spaces of S^2.

There are two important observations to make regarding 7.8.1. Firstly, for V a smooth projective curve, the group $\pi_1(V)^\wedge$ is <u>algebraically</u> defined, i.e. without using the complex topology or the concept of homotopy, and secondly that this approach opens the way to possible generalisations of the fundamental group for curves over arbitrary fields.

In Chapters 8, 10 and 11 we shall discuss smooth projective curves, and higher dimensional varieties endowed with the complex topology and use the familiar methods and tools of algebraic topology in order to obtain the desired results. But these arguments will in most cases be equally valid if one

replaces the classical cohomology theories by some algebraically defined
approximation, provided the relevant cohomology axioms are satisfied.   For
example, in Chapter 8, we are interested in $H^1(V,R)$ when V is a curve of
genus g.   Now $H^1(V,R)$ is the dual of $H_1(V,R)$ which is $H_1(V,Z) \otimes R$;
but $H_1(V,Z)$ is the abelianisation of $\pi_1(V)$ which we now know we can
approximate algebraically.   Unfortunately, the abelianisation does not
commute with completion.

But the inclusion $V \to J(V)$ of V into its Jacobian induces an
isomorphism on H_1 and since $J(V)$ is a topological group, $\pi_1(J(V))$ is
abelian and $\pi_1(J(V)) = H_1(J(V),Z)$.   So $H_1(V,Z) = \pi_1(J(V))$.   Provided we
can approximate the fundamental group of a smooth <u>variety</u> algebraically by
some generalisation of 7.8.1 (and we can) then we have an algebraic analogue
of $H_1(V)$.   We emphasise that in order to obtain $H_1(V)$ for a curve V, we
must move outside the category of curves and consider higher dimensional
varieties.

These remarks underly the relevance in later chapters of unramified
extensions and covering spaces (and their generalisations) and of Jacobians
(or more generally abelian varieties).   They also indicate the reasons one
prefers to work <u>algebraically</u> rather that <u>analytically</u>.

8 Elliptic curves

8.1   ELLIPTIC CURVES

If L is a lattice in C^n (i.e. a discrete free abelian subgroup of rank
2n) then the quotient group C^n/L inherits a complex manifold structure
such that the projection $C^n \to C^n/L$ is holomorphic.   With this structure,
C^n/L is a compact complex Lie group, and is called a <u>complex torus</u>.   As a
real Lie group, C^n/L is isomorphic to the 2n-dimensional torus T^{2n}.   If
n = 1,   then C/L is a compact Riemann surface of genus 1.

A surface $X \in Ob(\underline{RS}_0)$ can only be a topological group if it has genus
1,   since for genus 0 we have $X = S^2$ which is not a topological group,
(by the Lefchetz fixed point theorem[108]) while for $g \geq 2$, $\pi_1(X)$ is not
abelian[109].   Suppose $X \in Ob(\underline{RS}_0)$ has genus 1.   A choice of base point on
X determines (see §7.7) an isomorphism $X \to J(X)$ and hence a topological
group structure on X   We give a more direct proof of this fact.

<u>Definition 8.1.1</u>   <u>An elliptic curve</u> is a surface $X \in Ob(RS_0)$ of genus 1
with a chosen base point.

A complex torus C/L determines naturally an elliptic curve, based at 0.

For any connected Riemann surface X,   the group A(X) of isomorphisms
of X with itself is[110] a topological group with the compact-open topology.
It is in fact[111] a real Lie group, although not necessarily connected.   Let
X be an elliptic curve, and its universal covering space $X' \to X$ exhibits
X as the quotient space of X' by the group of covering transformations,
which form a discrete subgroup of A(X') naturally isomorphic to $\pi_1(X)$

which is free abelian of rank 2.

By the Riemann classification theorem[112], X' is isomorphic to precisely one of H, C or S^2. The automorphism groups of these three surfaces are well-known[113], and can all be presented as groups of Moebius transformations: $A(H) = PSL_2(R)$, $A(C) = \{(\begin{smallmatrix} a & b \\ 0 & 1 \end{smallmatrix}) \in GL_2(C)\}$ and $A(S^2) = PGL_2(C)$.

Although $PSL_2(R)$ contains free abelian subgroups of rank 2, none of them is discrete[114], and so $X' \neq H$, and hence $X' = C$, since clearly X' is non-compact. The elements of $A(C)$ are affine maps (of the complex Lie group C) so X can be identified with the quotient group C/L where $L \subset C$ is[115] the orbit of 0 under $\pi_1(X)$. Thus an elliptic curve is a complex torus.

The field M(X) of meromorphic functions on the elliptic curve $X = C/L$ can be identified with the field of meromorphic functions on C which are periodic rel L. Recall that a meromorphic function f on C is even or odd according as $f(z) = f(-z)$ or $f(z) = -f(-z)$. As usual any element of M(X) can be expressed uniquely as a sum of an odd and an even function.

There is one important element of M(X), the Weierstrass elliptic function which has formal Laurent series

$$\wp(z) = \wp_L(z) = \frac{1}{z^2} + \sum_{w \in L \setminus \{0\}} \left(\frac{1}{(z-w)^2} - \frac{1}{w^2} \right)$$

and derivative

$$\wp'(z) = \wp'_L(z) = 2 \sum_{w \in L} \frac{1}{(z-w)^3} .$$

The fact that $\wp$ and $\wp'$ converge uniformly on compact subsets of $C \setminus L$ follows from the next lemma.

__Lemma 8.1.2__   *For any lattice* $L \subset C$ *and* $r \in R$ *with* $r > 2$, *the series*

$$\sum_{w\in L\setminus\{0\}} w^{-r}$$

is absolutely convergent.

<u>Proof</u> Choose a basis w_1, w_2 for L, and let

$L_n = \{aw_1 + bw_2 \in L : \max\{|a|, |b|\} = n\}$. The series $\sum_{w\in L\setminus\{0\}} w^{-r}$ can be

rewritten $\sum_{n=1}^{\infty} \sum_{w\in L_n} w^{-r}$. If we choose d sufficiently small such that

$\{w \in L : |w| < d\} = \{0\}$, then for $w \in L_n$ we have $|w| > nd$. Since L_n

contains $8n$ points, we see that the series $\sum_{w\in L\setminus\{0\}} w^{-r}$ is dominated by

$\sum_{n=1}^{\infty} 8n\,(nd)^{-r} = 8d^{-r}\zeta(r-1)$.

The Laurent series expansions for $\wp$ and $\wp'$ show them to be meromorphic in C (with poles exactly the points of L) even and odd
respectively. Clearly $\wp'$ is periodic $\mathrm{rel}\,L$, and so for $w \in L$ there is
an element $C(w) \in C$ such that $\wp(z+w) - \wp(z) = C(w)$. Substituting
$z = -w/2$ (not a pole of $\wp$) we see that $C(w) = 0$ since $\wp(z) = \wp(-z)$, so
$\wp$ is periodic $\mathrm{rel}\,L$.

As a holomorphic function from X to S^2, $\wp$ has only 0 as a pole
and the Laurent series expansion shows this to be a double pole. The
ramification index at 0 is thus 2, so $\wp$ has valence 2. From the
Riemann–Hurwitz theorem (7.6.2), it follows that $\wp : X \to S^2$ is ramified at
four points. But since $\wp$ is even, if $z \in C/L$ the inverse image
$\wp^{-1}(\wp(z))$ contains two points if $z \neq -z$. The only points at which $\wp$ can
ramify are those for which $z = -z$. Since, as a group, X is the torus T^2,
there are four such points, the zero and the three elements e_1, e_2 and e_3
of order 2. Thus $\wp : X \to S^2$ is precisely the map identifying z with $-z$.
Similarly, $\wp' : X \to S^2$ has valence 3.

The map $\wp : X \to S^2$ exhibits $M(X)$ as a quadratic extension of $M(S^2)$

by 7.4.5.   Since $\wp'$ is odd, it does not lie in $M(S^2)$ and so generates $M(X)$ over $M(S^2)$.   We note that the automorphism $f \mapsto (z \mapsto f(-z))$ of $M(X)$ fixes $M(S^2)$ and generates the Galois group of the extension $M(X)/M(S^2)$.

<u>Theorem 8.1.3</u> *If $L \subset C$ is a lattice, then any meromorphic function f on C which is periodic rel L can be expressed uniquely in the form $a + b\wp'$, where a, b are rational functions of $\wp$.   Any even periodic function is a unique rational function of $\wp$.*

Now by 7.4.6, we know that there is a polynomial $F(t,u)$ of degree at most 3 in t and at most 2 in u such that $F(\wp,\wp') = 0$ in $M(X)$.   Since any polynomial in $\wp$ and $\wp'$ can only have poles at 0, we have only to find a polynomial which has no pole at 0.   Now $(\wp')^2 - 4\wp^3$ is an even periodic function with a double pole at 0.   So for suitable choice of $g_2, g_3 \in C$, the function $(\wp')^2 - 4\wp^3 - g_2\wp - g_3$ is periodic, has no poles, and vanishes at 0 so must be constant.   Hence $F(\wp, \wp') = 0$ where $F(t,u) = u^2 - 4t^3 - g_2 t - g_3$.   The values of g_2, g_3 can be explicitly described[116] as invariants of the lattice L.

If $z \in C$ is a point at which $\wp$ is ramified, then $\wp'$ vanishes at z. Thus the three complex numbers $\wp(e_i)$, $i = 1,2,3$ are the roots of the polynomial $4t^3 + g_2 t + g_3$, and so the discriminant $g_2^3 - 27g_3^2$ is non-zero. The polynomial $F(t,u)$ thus defines a smooth projective curve (see 6.7.1) $V = \{(z:x:y) \in P^2(C) : zy^2 - 4x^3 - g_2 xz^2 - g_3 z^3 = 0\}$, and a smooth affine curve $V_0 = V \cap C^2$.

The holomorphic map $z \mapsto (\wp(z), \wp'(z))$ from $C \backslash L/L$ to V_0 is an isomorphism, and this extends to an isomorphism $C/L \to V$ taking the point 0 to the point at infinity $(0:0:1)$, which we take as our base point for V', making it into an elliptic curve.   The analytic map $\wp : C/L \to S^2$ can be identified with the algebraic map $V \to P^1(C)$ taking $(z:x:y)$ to $(z:x)$

110

if $z \neq 0$, and $(0:1)$ if $z = 0$. We now show how the group structure on
V arises algebraically.

<u>Lemma 8.1.4</u> *For* $u,v,w \in C \backslash L,$ *the function*

$$f(u,v,w) = \det \begin{pmatrix} \wp(u) & \wp'(u) & 1 \\ \wp(v) & \wp'(v) & 1 \\ \wp(w) & \wp'(w) & 1 \end{pmatrix}$$

vanishes if $u + v + w = 0.$

<u>Proof</u> For fixed v, the meromorphic function $f(z,v,-z-v)$ of z is
periodic rel L with possible poles at $z = 0$ and $z = -w$. The series
expansions at these points show these not to be poles, so $f(z,v,-z-v)$ is
constant and in fact 0.

So if $u,v,w \in C/L$ have zero sum, their images in V lie on a line.
Conversely, any line in $P^2(C)$ meets V in three points (counting multiplicities) and the corresponding three points in C/L have zero sum. Thus
the group structure on V can be geometrically (or algebraically) defined as
follows: if P_1, P_2 are two points of V let P_3 be the (unique) third
point on V at which the line through P_1 and P_2 meets V (this line
being the tangent if $P_1 = P_2$) and let P_4 be the third point on the line
through P_3 and $(0:0:1)$. Then $P_4 = P_1 + P_2$.

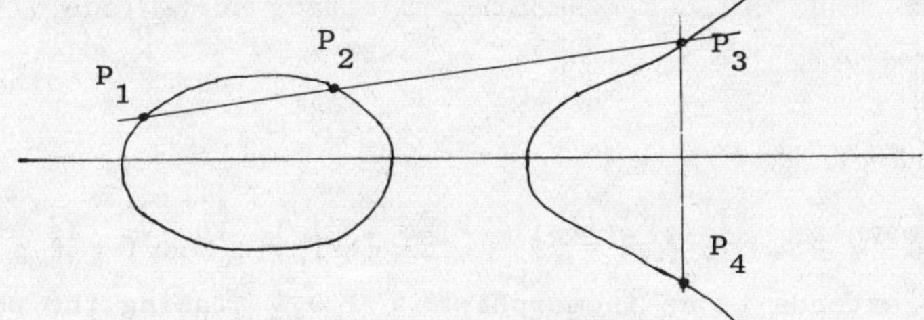

Now consider the three complex numbers $\wp(e_i)$, $i = 1,2,3$. The group
$A(C)$ acts 2-transitively on C, so we can apply an automorphism to C
taking $\wp(e_1)$, $\wp(e_2)$ to 0, 1. Let $a \in C \backslash \{0,1\}$ denote the image under

this automorphism of $\wp(e_3)$.   By a suitable change of variable in $P^2(C)$,
the curve V becomes the zero set $V(a)$ of the polynomial
$ZY^2 - X(X-1)(X-a) = g_a(X,Y,Z)$.   The value of a depends on the labelling of
the three points e_1, e_2, e_3.   A permutation of these indices would produce
an element in the S_3 orbit of a in $C\backslash\{0,1\}$ where S_3 acts in the usual
way as a group of Moebius transformations generated by the involutions
$a \mapsto 1 - a$, and $a \mapsto \dfrac{1}{a}$.

<u>Theorem 8.1.5</u>   *If* V *is an elliptic curve, then* $V \cong V(a)$ *for some*
$a \in C\backslash\{0,1\}$.   *In addition* $V(a) \cong V(a')$ *if and only if* $a = \sigma(a')$ *for*
some $\sigma \in S_3$.

<u>Remark 8.1.6</u>   The function $a \mapsto \prod_{\sigma \in S_3} (1+\sigma(a)) = \dfrac{(a+1)^2(2a-1)^2(a-2)}{a^2(a-1)^2}$ induces

a homeomorphism $h : C\backslash\{0,1\}/S_3 \to C$.   The group $PSL_2(Z) \subset PSL_2(R) = A(H)$
acts holomorphically on the upper half-plane H, and the kernel Γ_2 of the
reduction mod 2 map $PSL_2(Z) \to PSL_2(Z/2Z) = S_3$ is a free group of rank 2
(the modular group of level 2), isomorphic to $\pi_1(C\backslash\{0,1\})$.   There is a
commutative diagram

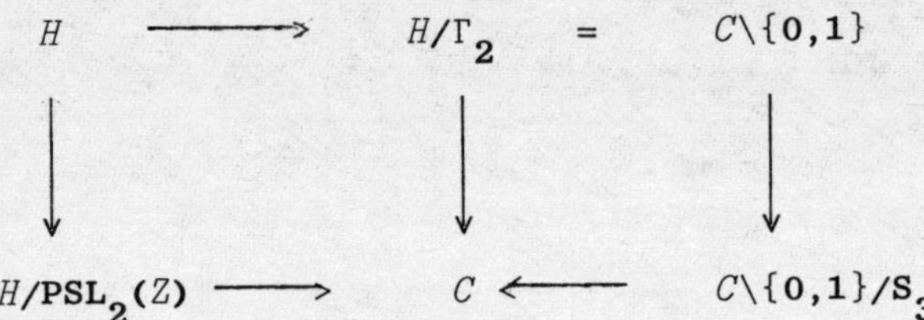

and the composite map $H \to H/PSL_2(Z) \to C$ is the elliptic modular function[117].

Let $X_i = C/L_i$ for $i = 1, 2$ be elliptic curves, and let $\mathrm{Hom}(X_1, X_2)$ be the group of holomorphic homomorphisms from X_1 to X_2. For $f \in \mathrm{Hom}(X_1, X_2)$, let $\tilde{f} : C \to C$ be the unique map between universal covering spaces lifting f and satisfying $\tilde{f}(0) = 0$. For any $a_1 \in L_1$ there is (by uniqueness of the lifting) an element $a_2 \in L_2$ such that $\tilde{f}(z+a_1) - \tilde{f}(z) = a_2$. Now $\tilde{f}$ is holomorphic, and $d\tilde{f}/dz$ is periodic rel L, so there is an $a \in C$ such that $\tilde{f}(z) = az$, and since $\tilde{f}(L_1) \subset L_2$ we have $aL_1 \subset L_2$. Conversely any element a of C for which $aL_1 \subset L_2$ induces a homomorphism from X_1 to X_2.

<u>Proposition 8.2.1</u> *The group* $\mathrm{Hom}(C/L, C/L_2)$ *is isomorphic to the subgroup* $\{a \in C : aL_1 \subset L_2\}$ *of* C. *The elliptic curves* $C/L_1, C/L_2$ *are isomorphic as complex Lie groups if and only if there is an* $a \in C^*$ *such that* $aL_1 = L_2$.

<u>Remark 8.2.2</u> An easy modification of the above shows that any holomorphic map between elliptic curves preserving the base point is a homomorphism, and hence any holomorphic map between elliptic curves of genus 1 is affine (with respect to their natural group structures). Hence two curves of genus 1 are isomorphic as Riemann surfaces if and only if they are isomorphic as complex Lie groups.

Now suppose $X = C/L$ is an elliptic curve. By 8.2.1 we can assume without loss of generality that L is a lattice of the form $L(\tau) = \{m + n\tau : m, n \in Z\}$ where $\tau \in H$. The ring $\mathrm{End}(X) = \mathrm{Hom}(X,X)$ can be identified with the subring $\{a \in C : aL \subset L\}$ of C and so is an integral domain of characteristic 0. Since $1 \in L$, the condition $aL \subset L$ implies that $a \in L$, so $\mathrm{End}(X)$ is a free abelian group of rank 1 or 2, the latter case occurring if and only if X admits endomorphisms other than integers, in which case we say X admits <u>complex multiplication</u>.

Suppose End(X) contains an element a which is not an integer, so that $a = m + n\tau$ for some m, n with $n \neq 0$.   Let $a\tau = m' + n'\tau \in L$. Elementary algebra shows that a is a root of $t^2 - (m+n')t + (m'n + mn')$. We have the following result immediately.

<u>Proposition 8.2.3</u> *Let* X *be the elliptic curve* $C/L(\tau)$ *with* $\tau \in H$.   *If* $\deg(Q(\tau)/Q) = 2$, *then* End(X) *is a free abelian group of rank 2 with every non-integer in* End(X) *a root of a monic integral quadratic polynomial with negative discriminant and* $Q \otimes$ End(X) $= Q(\tau)$.   *Otherwise* End(X) $= Z$.

Suppose $X = C/L(\tau)$ admits complex multiplication and that a $a = m + n\tau \in$ End(X) with $n \neq 0$.   With the above notation, the 'other root' of the polynomial $t^2 - (m+n')t + (m'n+mn')$ namely $a' = m + n' - a$ is also an element of End(X), and is the unique element of End(X) such that $aa' = m'n + mn'$ (since End(X) is an integral domain).   Now the integer $m'n + mn'$ is the index of the subgroup aL of L, or equivalently (since L/aL is isomorphic to Ker(a)), the order of Ker(a).   For any $a \in$ End(X) let $N(a) = [L:aL]$ if $a \neq 0$, and $N(0) = 0$.   The function $N: \text{End(X)} \to N$ is called the <u>norm</u>.   Clearly for $a \in Z \subset$ End(X) we have $N(a) = a^2$.

If $a = m + n\tau \in$ End(X) and $a\tau = m' + n'\tau$,   then $N(a+1) = (m+1)(n'+1) - m'n = N(a) + m + n + 1$ so that a is a root of the polynomial

$$P_a(t) = t^2 - (N(a+1) - N(a) - 1)t + N(a)$$

while $a' = N(a+1) - N(a) - 1 - a$.   This equation expresses the element a' in terms of a and the function N.   The following result is trivial.

<u>Lemma 8.2.4</u> *The function* $a \mapsto a'$ *from* End(X) *to itself is a ring automorphism.*

114

<u>Remark 8.2.5</u>   Of course, once we have identified End(X) with a subring of C the element a′ is no more than the complex conjugate $\bar{a}$ of a.   The reason for describing a′ in terms of the norm N is that this approach is algebraic and intrinsic to the elliptic curve X.   For N(a) can be algebraically described as the number of points of intersection (counting multiplicities) of the curve $X \times \{0\}$ on $X \times X$ with the graph $\{(x,a(x)) : x \in X\}$ of the endomorphism a, and 8.2.4 has an algebraic proof. This will be relevant in the next section.

   We now give the Hasse proof of a complex analogue of the Riemann Hypothesis for elliptic curves.

<u>Proposition 8.2.6</u>   *Let* X *be a complex elliptic curve and* $\pi : X \to X$ *a holomorphic homomorphism whose kernel is finite of order* q > 1.   *The number* ν_n *of fixed points of* π^n *is finite and*

$$\exp \sum_{n=1}^{\infty} \nu_n t^n/n \;=\; P_\pi(t)/(1-t)(1-qt).$$

Moreover the roots of the quadratic polynomial $P_\pi(t)$ *have absolute value* $q^{-\frac{1}{2}}$.

<u>Proof</u>   The element $1 - \pi^n \in \text{End}(X)$ is non-zero since π is not an isomorphism, and so $\nu_n = |\text{Ker}(1 - \pi^n)| = N(1 - \pi^n)$

$$= (1 - \pi^n)(1 - \pi'^n) \quad (\text{in End}(X)).$$

Hence

$$\exp \sum_{n=1}^{\infty} \nu_n t^n/n \;=\; \exp \sum_{n=1}^{\infty} (t^n - (\pi t)^n - (\pi' t)^n + (\pi \pi' t)^n)/n$$

$$= P_\pi(t)/(1-t)(1-qt), \quad \text{using 8.2.4.}$$

   The result follows since $N(\pi) = \pi \pi' = q.$

Let K be q field of transcendence degree 1 over the finite field F_q of
order q, of genus 1. The associated smooth model V for K is called an
<u>elliptic curve</u> over F_q.   We describe in this section how Hasse obtained
the form of the zeta-function for K on which 8.2.6 has been based, by
indicating where the proof of 8.2.6 must be modified.

Firstly, V is an abelian group with the operations defined algebraically
and the Frobenius map $\pi: V \rightarrow V$ is a homomorphism.   If End(V) denotes the
ring of endomorphisms of V which are defined algeb aically, then End(V) has
no zero divisors and has characteristic zero although it may be non-
commutative[119].   For each a $\in$ End(V) the subgroup Ker(a) $\subseteq$ End(V) is finite,
and the intersection number N(a) of V x $\{0\}$ with the graph of a on
V x V can be defined.[120]   Although $\pi: V \rightarrow V$ is an isomorphism, the zero of
V is a q-fold zero, so $N(\pi) = q$.   Each element a $\in$ End(V) is a root of the
polynomial $t^2 - (N(a+1) - N(a) - 1)t + N(a)$, and so the element
$\bar{a} = N(a+1) - N(a) - 1 - a$ belongs to End(V) and is the unique element
satisfying $a\bar{a} = \bar{a}a = N(a)$.   The function $a \rightarrow \bar{a}$ is an autiautomorphism
of End(V), called <u>conjugation</u>.

Hence the ring $E_Q = Q \otimes End(V)$ is a division algebra over Q in which
each non-rational element generates an imaginary quadratic extension of Q .
Such division algebras are classified[121], so E_Q is isomorphic to one of
(i) Q , (ii) an imaginary quadratic number field, or (iii) a quaternionic
algebra $Q(a,b)$ with a^2, b^2 negative rational numbers and ab = -ba.

In any case, conjugation in End(V) corresponds to conjugation (in the
usual sense) in these division algebras.

Thus the proof of 8.2.6 can now be taken verbatim, since for each n $\in N$,
the endomorphism $1 - \pi^n$ of V has no multiple zeros, so $|Ker(1-\pi^n)| = N(1-\pi^n)$.

This gives the proof of the Weil conjectures for curves of genus 1, due to Hasse.

<u>Theorem 8.3.1</u> *If* K *is a field of transcendence degree* 1 *over* F_q *and genus* 1 *, then* $Z(t,K) = 1 + at + qt^2/(1 - t)(1 - qt)$ *for some integer* a.

The precise form of this expression shows that in particular $Z(t,K)$ is rational, satisfies the functional equation, and that its zeros have absolute value $q^{-1/2}$.

<u>Example 8.3.2</u> Let us return to example 3.5.6. The field K has transcendence degree 1 over F_7 and has genus 1 by (3.3.3). Consider the smooth projective model V (see 6.7.1) in $P^2(F_7)$ defined by the polynomial

$$g(Z_2,X_2,Y_2) = Z_2Y_2^2 - (5X_2^3 + 2X_2^2Z_2 + 3X_2Z_2^2 + 3Z_2^3).$$

There is only one point on the line $Z_2 = 0$, namely the point $(0 : 0 : 1)$. Thus the points of $V \cap P^2(F_7)$ consist of this point together with the solutions of $g(1,X_2,Y_2) = 0$ over F_7, namely the points $(2,1)$, $(2,6)$, $(3,2)$, $(3,5)$, $(5,0)$, $(6,2)$ and $(6,5)$, and so $\nu_1 = 8$. Equating coefficients in $\exp \Sigma \nu_n t^n/n = 1 + at + 7t^2/(1 - t)(1 - 7t)$ we see that $a = 0$.

This curve is smooth so we see immediately that the Z-function of the extension $K/F_7(X_2)$ is

$$Z(t,K/F_7(X_2)) = 1 + 7t^2/(1 - 7t),$$

since removing from V the point $(0 : 0 : 1)$ which lies in $P^2(F_7)$ we obtain a smooth affine curve whose points corredpond to the spots on K which are not infinite on $F_7(X_2)$, using 6.5.1, 8.3.1 and the results of 5.5.

8.4   APPENDIX: ISOGENIES

We remark briefly that much of the previous sections of this chapter is valid in a generalised form for complex tori C^n/L for $n \geq 1$, provided one considers not just homomorphisms between such tori but isogenies.

Definition 8.4.1   A holomorphic homomorphism $a: C^n/L \to C^m/L'$ is an
isogeny if $m = n$, and a is surjective with finite kernel.

An isogeny into the torus C^n/L is thus precisely a finite covering
space.   Now the holomorphic endomorphisms of $X = C^n/L$ can be identified
with the matrices $M \in M_n(C)$ for which $M(L) \subset L$.   Such a matrix M
corresponds to an isogeny if and only if it belongs to $GL_n(C)$.   More
directly, $a: X \to X$ is an isogeny if and only if a is invertible in
$Q \otimes End(X)$.

Suppose $a: X \to X$ is an isogeny which corresponds to the linear map
$\tilde{a}: C^n \to C^n$.   The subgroup $a(L)$ of L has finite index, say $N(a)$,   equal
to the determinant of the R-linear map $a_R = 1 \otimes \tilde{a} : R \otimes L \to R \otimes L$.

If $\pi: X \to X$ is an isogeny such that each power π^n is also an isogeny
(in other words the eigenvalues of $\tilde{\pi}$ are not roots of unity), then for each
n the element $1 - \pi^n$ of $End(X)$ is an isogeny, and
$\nu_n = \left| Fix(\pi^n) \right| = det(1 - \pi_R)$.   Now if f is an endomorphism of the finite
dimensional vector space V over some field F and we write $\chi_t(f)$ for
$det(1 - tf)$,   then the following lemma is an easy exercise in canonical forms.

Lemma 8.4.2   *With the above notation, we have*

$$t d/dt \left[\log\left(\prod_{i=0}^{dim V} (\chi_t(\lambda^i f))^{(-1)^{i+1}} \right) \right] = \sum_{m=v}^{\infty} det(1 - f^m) t^m.$$

Using this lemma and adapting the proof of 8.2.6, we have the following.

Theorem 8.4.3   *If* V *is a complex torus of dimension* m *and* $\pi: V \to V$ *is
an isogeny whose eigenvalues are not roots of unity, then the number* ν_n *of
fixed points of* π^n *is finite for each* n *and we have*

$$\exp \sum \nu_n t^n/n \;=\; \prod_{i=0}^{m} [\chi_t(\lambda^i \pi)^{(-1)^{i+1}}].$$

In particular, $\exp \sum \nu_n t^n/n$ *is a rational function of* t.

9 Varieties

Classical algebraic geometry deals with the solution of polynomial equations over the complex numbers.   More generally one is interested in solutions of polynomials over an arbitrary algebraically closed field, possibly of non-zero characteristic.   These polynomials may have coefficients in a proper subfield k of K which may or may not be algebraically closed.   This leads to the concept of a K/k-variety (or relative variety).   A precise description of such an object is rather technical, and uses the language of sheaf theory in order to piece together affine varieties.   We need this to discuss projective varieties, which have the property of being 'complete'; two lines in A^2 may not intersect, but in P^2 any two distinct lines meet in a single point.

Throughout this chapter, k will be any field and K/k an extension with K algebraically closed.   The symbols A^n, P^n denote the affine and projective spaces over K.

9.1   AFFINE VARIETIES

Let $k[A^n]$ denote the ring of polynomials over k in n variables, whose elements we consider as functions from A^n to K.   The ring $k[A^n]$ is Noetherian by Hilbert's basis theorem[122].   If J is an ideal of $k[A^n]$, let $V(J) = \{x \in A^n : f(x) = 0 \text{ for all } f \in J\}$.   Since J is finitely generated, the set $V(J)$ is the set of common zeros of a finite set of polynomials.   Conversely the set of common zeros of a finite set of polynomials is equal to $V(J)$ where J is the ideal they generate.

It is easy to see that $V(\Sigma_\alpha J_\alpha) = \cap_\alpha V(J_\alpha)$ and $V(J_1 \cap J_2) = V(J_1) \cup V(J_2)$ so that A^n can be given a unique topology with the sets $V(J)$ for $J \in k[A^n]$ as the closed sets. This topology, and the topology it induces on any subset, is called the <u>Zariski topology</u>, or the <u>k-topology</u> to emphasise its dependence on $k \subset K$.

<u>Example 9.1.1</u>   If $k = K$, then the points of A^n are closed. The non-trivial closed sets in A^1 are the finite sets, and in A^2 are either finite or points of a curve. In particular the topology is non-Hausdorff, and the topology on A^2 is not the usual product topology on $A^1 \times A^1$. If $k \neq K$, then in general points are not closed.

Suppose $S \subset A^n$ is a closed set. Let $I(S) = \{f \in k[A^n] : f(x) = 0 \text{ for all } x \in S\}$, so that $I(S)$ is an ideal of $k[A^n]$, with $VI(S) = S$. Recall that for any ideal J of $k[A^n]$ its radical $\sqrt{J} = \{f \in k[A^n] : f^r \in I \text{ for some } r \in N\}$ is an ideal containing J. Clearly $V(\sqrt{J}) = V(J)$. The important result is Hilbert's Nullstellensatz[123].

<u>Theorem 9.1.2</u>   *If* J *is an ideal of* $k[A^n]$ *then* $IV(J) = \sqrt{J}$.

Thus the function V is an order-reversing isomorphism from the lattice of ideals in $k[A^n]$ to the lattice of closed subsets of A^n. A closed subspace of A^n is called an <u>affine</u> K/k-<u>variety</u> (or an <u>affine</u> K-<u>variety</u> if $k = K$); by abuse of notation we use the word variety generically.

Recall that a topological space X is <u>Noetherian</u> if the open sets satisfy the ascending chain condition, and that X is <u>irreducible</u> if it satisfies one of the equivalent conditions that X is not the union of two proper closed subsets, any two proper open sets intersect non-trivially and any open subset is dense. Noetherian spaces are quasi-compact.

Now suppose that $V \subset A^n$ is an affine variety. By virtue of the correspondence between the closed subsets of V and the ideals of $k[A^n]$ containing $I(V)$, the fact that $k[A^n]$ is Noetherian implies that V is Noetherian, and that V is irreducible if and only if $I(V)$ is a prime ideal. In the ring $k[A^n]$ every ideal J has a minimal primary decomposition[124] (corresponding to factorisation of ideals in rings of integers of a number field), that is to say there are primary ideals $J_1,\ldots,J_r$ such that $J = \cap J_i$, the radicals $\sqrt{J_i}$ are distinct, and $J_i \nsubseteq J_j$ for $i \neq J$. Now since $V(J) = V(\cap J_i) = \cup V(J_i) = V(\sqrt{J_i})$ we see that each affine variety V can be expressed uniquely as a union $V = V_1 \cup \ldots \cup V_r$ of irreducible varieties such that $V_i \nsubseteq V_j$ for $i \neq j$. These pieces V_i of V are called the <u>irreducible components</u> of V.

9.2   REGULAR FUNCTIONS

Suppose $V \subset A^n$ is an affine K/k-variety. A function $\tilde{f} : V \to K$ is <u>regular</u> (over k) if there is an element $f \in k[A^n]$ such that $f(v) = \tilde{f}(v)$ for all $v \in V$. Such a function is continuous and the set $k[V]$ of regular functions on V is a ring under pointwise operations, isomorphic to $k[A^n]/I(V)$. The ring $k[V]$ is an integral domain if and only if V is irreducible, in which case the elements of the field of fractions $k(V)$ of $k[V]$ are called <u>rational functions</u> on V (over k). The transcendence degree of $k(V)$ over k is finite, and is called the <u>dimension</u> of V, written $\dim(V)$. If V is not irreducible, we put $\dim(V) = \max \dim(V_i)$ where $V_1,\ldots,V_r$ are the irreducible components of V. If $\dim(V) = 1$, we say V is a curve. If W is a closed subspace of V then $\dim(V)-\dim(W)$ is called the <u>codimension</u> (of W in V).

Each point $v \subset V$ determines an evaluation homomorphism $e_v: k[V] \to K$, $f \mapsto f(v)$ and since $\text{Im}(e_v)$ is an integral domain the kernel p_v of e_v is a prime ideal. It may be maximal (e.g. if $k = K$, or if V is a curve as in §6.4) or may not.

Example 9.2.1   If $k = R$ and $K = C|t|$ and V is the curve $\{(x,y) \in A^2: x^2 + y^2 = 1\}$, then the point $v = (\cos t, \sin t)$ lies on V, and the image of e_v in K is the R-subalgebra generated by $\cos t$ and $\sin t$ which is not a field, since $(\cos t)^{-1}$ is not a polynomial function of $\cos t$ and $\sin t$.

Let $O_{V,v}$ denote the localisation of the ring $k[V]$ at the prime ideal p_v. If V is irreducible, then $O_{V,v}$ is the subring of $k(V)$ consisting of functions of the form f/g with $g(v) \neq 0$. The kernel of the evaluation map $e_v: O_{V,v} \to K$ is the unique maximal ideal m_v of this local ring. The quotient field $k(v) = O_{V,v}/m_v$ is the subfield of K generated by the coefficients of v, and is called the residue field of V at v.

Example 9.2.2   If $V \subset A^2$ is the curve corresponding to an extension $E/k(X)$, then $k(V) = E$ and $O_{V,v}$ is the valuation ring corresponding to the p_v-adic spot on E, while $k(v)$ is the corresponding residue field.

Definition 9.2.3   If V is an irreducible affine K/k-variety, and element $f \in k(V)$ is regular at $v \in V$ if $f \in O_{V,v}$.

For any open set $U \subset V$, let $O_V(U)$ be the subring $\underset{v \in U}{\cap}\, O_{V,v} = \{f \in k(V): f$ is regular at v for each $v \in V\}$. The functor $O_V: \underline{\text{Ouv}(V)} \to \underline{\text{k-alg}}$ from the category of open subsets of V and inclusion maps to the category of k-algebras is a sheaf, and moreover we have $k[V] = O_V(V)$, $k(V) = \underset{\to}{\lim}\{O_V(U): U \epsilon \text{Ouv}(V)\}$ and $O_{V,v} = \underset{\to}{\lim}\{O_V(U): U \in \text{Ouv}(V), v \in U\}$. This should be compared with §7.4.

If $V \subset A^n$ and $W \subset A^m$ are affine K/k-varieties, a function $f: V \to W$ is _regular_ if each component $f_i : V \to K$ is regular. Such a map induces a homomorphism from $k[W]$ to $k[V]$ by composition.

If $V \subset A^n$ and $W \subset A^m$ are affine varieties, their cartesian product $V \times W \subset A^{n+m}$ is the set of zeros of some ideal, so is itself a variety. The projection maps from $V \times W$ to V and W are regular, and so the category of affine K/k-varieties and regular maps has products. We note that the Zariski topology on $V \times W$ is not in general the product topology.

Now $V \longmapsto k[V]$ is a contravariant functor from the category $\underline{V(K/k)}$ of irreducible affine varieties and regular maps, to the category of finitely-generated integral domains over k and k-homomorphisms.

This functor is an equivalence of categories. For if E is a finitely generated integral domain over k, then there exists a surjective homomorphism $f: .k[A^n] \to E$ for some n; the affine variety $V(\ker f) \subset A^n$ defines the inverse construction.

More generally, the category of K/k-affine varieties and k-regular maps is equivalent to the category of k-algebras with no nilpotent elements, and k-homomorphisms. This latter category is independent of K.

9.3   POINTS AND PRIME IDEALS

<u>Definition 9.3.1</u> For a ring R the set $\mathrm{spec}(R)$ of prime ideals of R is called the _spectrum_ of R.

If $V \subset A^n$ is an irreducible affine K/k-variety then $p: V \longmapsto p_r$ is a function from V to $\mathrm{spec}(k[V])$, which in general is neither injective nor surjective. There are two special cases when one can say more about this function, and these two cases occur frequently in algebraic geometry, usually built into the axioms of the extension K/k.

For $p \in \text{spec}(k[V])$ the inverse image p' of p under the projection $k[A^n] \to k[V]$ is a prime ideal containing I(V) and so W = V(p') is an irreducible variety contained in V of codimension 1.   In order that $p = p_v$ for some point $v \in V$ it is necessary and sufficient that k(W) embeds in K so that p is the kernel of the composite

$$k[V] \to k[V]/p = k[W] \hookrightarrow k(W) \hookrightarrow K.$$

This will not be possible unless tr.deg(K/k) is at least dim(V)-1.   If V is a curve, this is no restriction.   But in order to work with varieties of arbitrarily high dimensions, it would be natural to postulate that K has infinite transcendence degree over k, i.e. K is a <u>universal domain</u>[125] for k.   Such extensions exist, for example K could be the algebraic closure of the field of rational functions over k in infinitely many variables.   If K is a universal domain over k, then the natural map $V \to \text{spec}(k[V])$ is surjective.   The condition that two points $v_1, v_2 \in V$ determine the same prime ideal is that any regular function on V should vanish on v_1 if and only if it vanishes on v_2, that is to say the sets $\{v_1\}$ and $\{v_2\}$ have the same closure (in the Zariski topology) as that v_1, v_2 are <u>specialisations</u>[126] of each other.

This is the first case.   The second case is the other extreme when k = K.   An immediate corollary of 9.1.2 is Hilbert's Weak Nullstellensatz.

<u>Theorem 9.3.2</u>   *If* K *is an algebraically closed field, then the function* $p: A^n \to \text{max}(K[A^n])$ *is a bijection.*

It follows that if V is an irreducible affine K-variety, then the function $p: V \to \text{max}(K[V])$ is also a bijection.

9.4   TANGENT SPACES AND SIMPLE POINTS

In this section we assume k = K and that V is an affine K-variety.   In

124

order to define tangency, we must make precise the meaning of multiple intersection of a line with V, or more generally of intersection numbers.   We do this exactly as in the classical case.   Suppose v is the origin for simplicity.   If L is a line through 0, then $L = \{ta: t \in K\}$ for any point $a \in L \setminus \{0\}$.   The <u>intersection number</u> $I(V,L;v)$ is the largest power of t which divides all the polynomials $f(ta) \in K[t]$ for $f \in I(V)$.   This definition is independent of the choice of a.   We say that L <u>touches</u> V at v, or that L is a <u>tangent</u> to V at v if $I(V,L;v) > 1$.   The union of all tangents at v is a vector subspace

<u>Definition 9.4.1</u>   The point v of the variety V is <u>simple</u> if $\dim T_v = \dim(V)$, otherwise it is <u>singular</u>.

This definition is straight-forward and familiar, but it has the disadvantage (or so it seems) of depending on how V sits inside A^n.   We now show how to give an intrinsic definition.

Again suppose $v = 0$ is the origin, and let T be the tangent space in the above sense.   If $g \in k[A^n]$ the partial derivative $\partial g / \partial X_i$ also belongs to $k[A^n]$ and can be evaluated at any point a of A^n, and we write $\partial g / \partial x_i \big|_a$ for its value.

Now for $a \in T$ and $f \in I(V)$ the element $\Sigma \partial g / \partial X_i \big|_0 a_i$ of K is zero and so the function from the maximal ideal m of $k[V]$ of functions vanishing at v, to the dual T* of the tangent space of T, which sends $f \in m$ to the functional $a \longmapsto \Sigma \partial \tilde{f} / \partial X_i \big|_0 a_i$ for some $\tilde{f} \in k[A^n]$ representing f, is well-defined.   This K-homomorphism vanishes on m^2 and induces an isomorphism $m/m^2 \to T*$.   This enables us to make the new definition.

<u>Definition 9.4.2</u>   The <u>tangent space</u> T_v at the point v of the affine K-variety V is the dual of the $K = 0_{V,v}/m_v$ vector space m_v/m_v^2, where m_v is the unique maximal ideal of $0_{V,v}$.

The definition of simple point (9.4.1) still applies.   Suppose V is
the zero-set of the polynomials $f_1, \ldots f_r \in K[A^n]$.   It is easy to see that
a point v of V is simple if and only if the Jacobian matrix $\| \partial f_i / \partial X_j \|$
has maximum rank at v.   If its rank is not maximum, then some minor of the
matrix vanishes.   The set of common zeros of the minors of the Jacobian is a
variety which intersects V in precisely the singular points of V, which
thus constitute a subvariety of codimension at least 1.   This is the
analogue of 6.2.1.

In a similar way, we say that the point $v \in V$ is <u>normal</u> if the ring $O_{V,v}$
is integrally closed (in its field of fractions, which will be K(V) if V is
irreducible), and that V is normal if each point is <u>normal</u>.   Since integral
closure is a local property ,   an irreducible variety V is normal if and
only if $k[V]$ is integrally closed.   It can be shown that a simple point is
normal[127] (so smooth implies normal) and that for a normal variety V the set
of singular points has[128] codimension at least 2.   This explains why for
curves integral closure and smoothness are equivalent,   and indicates why for
varieties of higher dimensions smoothness is a natural generalisation of
integral closure.

9.5   SHEAVES

Let $f: Y \to X$ be any continuous map between topological spaces.   For U open
in X, let $\Gamma_Y(U) = \{s: U \to Y$ such that $fs = 1\}$ be the set of continuous
sections of f over U.   For $U' \subset U$ there is a restriction map from $\Gamma_Y(U)$ to
$\Gamma_Y(U')$.   If U is covered by the open sets $\{U_i : i \in I\}$ and $s, s' \in \Gamma_Y(U)$
are two elements whose restrictions to U_i coincide for each $i \in I$, then
s = s'.   Conversely. if for each $i \in I$ we have an element $s_i \in \Gamma_Y(U_i)$
such that s_i agrees with s_j on $U_i \cap U_j$ then there is a unique element
$s \in \Gamma_Y(U)$ whose restriction to U_i is s_i for each i.   That is to say,

the sequence of sets

$$* \longrightarrow \Gamma_Y(U) \longrightarrow \prod_i \Gamma_Y(U_i) \quad \overset{>}{\underset{>}{\geq}} \; \prod_{i,j} \Gamma_Y(U_i \cap U_j)$$

is exact, where * denotes a terminal object in the category of sets, i.e. a one-point set.

   Recall that $\underline{Ouv(X)}$ denotes the category of open subsets of X and inclusion maps.   A contravariant functor F: $\underline{Ouv(X)} \longrightarrow \underline{Sets}$ is called a presheaf (of sets) on X, and if F satisfies the condition that for any open covering $\{U_i : i \in I\}$ of U the sequence

$$* \longrightarrow F(U) \longrightarrow \prod_i F(U_i) \quad \overset{>}{\underset{>}{\geq}} \; \prod_{i,j} F(U_i \cap U_j)$$

is exact, the F is called a $\underline{sheaf}$.   More generally, replacing $\underline{Sets}$ by any category $\underline{C}$ with suitable limits, we have a definition of $\underline{presheaves}$ and $\underline{sheaves}$ on $\cdot$ X $\underline{with\ values\ in\ C}$.   By a $\underline{morphism}$ of presheaves (or sheaves) we mean a natural transformation of functors.   We write $\underline{Sh(X)}$ for the category of sheaves on X.   The functor Γ_Y described above is a sheaf, called the $\underline{sheaf\ of\ sections}$ of f: Y → X.   The sheaf of sections of the projection map X × F → X is called the $\underline{constant}$ or $\underline{trivial}$ sheaf at F. For example, if R is a topological ring, and R(-) is the constant sheaf at R, then R(U) is the R-algebra of continuous functions from U to R. The sheaves which concern us initially are all subsheaves of R(-) for suitable R.

Example 9.5.1   For U an open subset of R^n and $r \in N \cup \{\infty, \omega\}$, let $D^r(U) = \{f \in R(U): f$ is r-times differentiable$\}$.   The sheaf D^r is called the sheaf of $\underline{r\text{-times differentiable functions}}$.

Example 9.5.2   If U is open in C^n, let H(U) = $\{f \in C(U): f$ is holomorphic$\}$.   The sheaf H is called the sheaf of $\underline{holomorphic\ functions}$.

<u>Example 9.5.3</u>   The field K is a topological ring with the k-topology.
If V is an affine K/k-variety (with the k-topology) and U is open in V,
then $O_V(U)$ is a subring of K(U) and O_V: <u>Ouv(V)</u> $\to$ <u>k-alg</u> is a sheaf, called
the sheaf of <u>k-regular functions</u>.

  If f: X $\to$ Y is a continuous map, then f induces a functor
f^{-1}: <u>Ouv(Y)</u> $\to$ <u>Ouv(X)</u> taking U to $f^{-1}(U)$, and so any presheaf (sheaf) F
on X induces by composition a presheaf (sheaf) $f_*(F) = F \cdot f^{-1}$ on Y.

  If F: <u>Ouv(X)</u> $\to$ <u>C</u> is a presheaf, then the limit
$\varinjlim\{F(U): U \in Ouv(X), x \in U\}$ (taken in the category <u>C</u>) is called the <u>stalk</u>
of F at x, and is denoted by F_x.   If F is a presheaf of sets and
$s \in F(U)$, we write s(x) for the element of F_x determined by s .   If F
is a presheaf of sets, let $e(F) = \coprod_{x \in X} F_x$ be the disjoint union of the stalks,
and let f: e(F) $\to$ X be the projection taking F_x to the point x.   We
give e(F) the topology with basis of open sets $\{N(U,s): U \in Ouv(X), s \in F(U)\}$
where $N(U,s) = \{s(x) \in e(F): x \in U\}$.   With this topology, the map
f: e(F) $\to$ X is a local homeomorphism, and is called the <u>espace etale</u>
associated to F.   In general, the space e(F) is not Hausdorff.

  There is a natural morphism from F to the sheaf of sections of
f: e(F) $\to$ X which is an isomorphism[129] if and only if F is a sheaf.   This
construction defines a functor from the category of presheaves on X to the
category of sheaves on X, adjoint to the inclusion.

  For any category <u>C</u> and X $\in$ Ob(<u>C</u>) the category <u>C/X</u> has objects maps
f: Y $\to$ X in <u>C</u>, and morphisms from f: Y $\to$ X to f': Y' $\to$ X those maps
g: Y $\to$ Y' in <u>C</u> such that f'g = f.   The category <u>C/X</u> is called the
<u>category of objects over X</u>.   For example, if <u>C</u> = <u>Rings</u> then for any ring
R, the category <u>C/R</u> is the opposite of the category of R-algebras.   Now
the sheaf of sections is a functor from <u>Top/X</u> to <u>Sh(X)</u>.   Let $\underline{X}_{1h}$ be the

128

full subcategory of $\underline{Top/X}$ whose objects are local homeomorphisms $f: Y \to X$.

The following proposition gives an alternative approach to sheaves[130].

Proposition 9.5.4 *The functor from $\underline{Sh(X)}$ to $\underline{X}_{1h}$ which associates to a sheaf its éspace étale is an equivalence with inverse given by the sheaf of sections functor.*

If $f: X \to Y$ is a continuous map and F is a sheaf on Y, then the pull-back $e(F)x_Y X \to X$ is a local homeomorphism, and its sheaf of sections is denoted by $f^*(F)$. If f is the inclusion of a subspace, we write $F|_X$ for $f^*(F)$.

Example 9.5.5 If X is a subspace of R^n and $U \subset X$ is open, the $D^r|_X(U)$ is the ring of continuous functions on U which are r-times differentiable at each point of U. Recall that $f: U \to R$ is r-times differentiable at $x \in U$ if and only if there is a neighbourhood U_x of x in R^n and an element $g \in D^r(U_x)$ such that $g(y) = f(y)$ for all $y \in U_x \cap U$.

Example 9.5.6 If $V \subset A^n$ is an affine K/k-variety, then $O_V = O_{A^n}|_V$.

Example 9.5.7 If U is open in X and F is a sheaf on X, then $F|_U$ is the restriction of F to the subcategory $\underline{Ouv(U)}$ of $\underline{Ouv(X)}$.

9.6   RINGED SPACES

In this section we use sheaves to piece together affine varieties to obtain more general varieties. We are particularly interested in the projective varieties (section 9.7) and the sheaf-theoretic approach includes the definition of maps between such objects which is otherwise equally technical. At the same time this construction in more generality enables us to define abstract manifolds (topological, differentiable, complex etc.) from the 'affine' pieces. We shall also need this construction later when we consider schemes.

Let R be a ring. The category R-Top has objects pairs (X,F) where
X is a topological space, and F is a sheaf of R-algebras on X. Objects
of $\mathbb{Z}$-top are called ringed spaces. We refer to X as the base space of
(X,F), and to F as the structure sheaf. A morphism from (X,F) to
(X',F') in R-Top is a pair (f,α) where f: X $\to$ X' is a continuous map,
between the base spaces, and α: F' $\to$ f$_*$(F) is a morphism between

sheaves of R-algebras over X'. The composition of (f,α) with
(f',α') : (X',F') $\to$ (X",F") in R-Top is the morphism (f'f,f'$_*$(α)α').
The forgetful functor from R-Top to Top ((X,F)$\longmapsto$X) is denoted by Φ.

Remark 9.6.1   In relation to a particular object (X,F) in R-Top$_0$, if Y
is a subspace of X we shall often use the symbol Y also to denote
(Y,F$|_Y$).

A morphism (f,α): (X,F) $\to$ (Y,G) induces in the limit an R-homomorphism
α_x: G$_{f(x)}$ $\to$ F$_x$ between stalks. The category in which we shall work is the
subcategory R-Top$_0$ of R-Top whose objects are those (X,F) for which F$_x$
is a local ring for each x $\in$ X, and whose morphisms from (X,F) to (Y,G)
are local homomorphisms on each stalk, i.e. for which $\alpha_x^{-1}(m_x) = m_{f(x)}$ for
each x $\in$ X, where m$_x$ denotes the maximal ideal in the stalk at x. The
quotient ring F(x) of the stalk F$_x$ by its maximal ideal m$_x$ is called
residue field of F at x.

The vector space $T_x^* = m_x/m_x^2$ over the residue field F$_x$/m$_x$ is called
the cotangent space at x, and its dual $T_x = (T_x^*)^*$ is the tangent space.
A morphism in R-Top$_0$ induces homomorphisms between residue fields, tangent
and cotangent spaces in the obvious way.

Let $\underline{C}$ be a category, and $A:\underline{C} \to \underline{R\text{-Top}}_0$ a functor satisfying

(9.6.2) A *is full and faithful*.

Any object in $\underline{R\text{-Top}}_0$ which is isomorphic to $A(c)$ for some $c \in \mathrm{Ob}(\underline{C})$ is called <u>affine</u> (with respect to A). An object (X,F) of $\underline{R\text{-Top}}_0$ is <u>locally affine</u> if for each $x \in X$ there is an open set U containing x such that $(U,F|_U)$ is affine. The full subcategory of $\underline{R\text{-Top}}_0$ of locally affine objects is denoted by $\underline{\mathrm{la}(A)}$.

<u>Example 9.6.3</u> Let $\underline{R}$ be the category whose objects are open subsets of $\underline{R}^n$ for some n, and whose morphisms from U to V are differentiable functions from U to V. For any $r \in N \cup \{\infty,\omega\}$, the functor $D^r: \underline{R} \to \underline{R\text{-Top}}_0$ taking U to $(U,D^r|_U)$ (see 9.5.2) satisfies (9.6.2). The objects of $\underline{\mathrm{la}(D^r)}$ are (not necessarily Hausdorff or separable) r-times differentiable manifolds, and the morphisms are r-times differentiable maps.

<u>Example 9.6.4</u> Let $\underline{C}$ be the category of the open subsets of $\underline{C}^n$ for some n with morphisms holomorphic maps. The functor $H:\underline{C} \to \underline{C\text{-Top}}_0$ taking U to $(U,H|_U)$ satisfies (9.6.2) are objects of $\underline{\mathrm{la}(H)}$ are (not necessarily Hausdorff or separable) complex manifolds, and the morphisms are holomorphic maps.

For our extension K/k, let $\underline{V(K/k)}$ be the category whose objects are irreducible affine varieties and whose morphisms are regular maps (see §9.2). The functor $V \mapsto (V,O_V)$ from $\underline{V(K/k)}$ to $\underline{k\text{-Top}}$ satisfies (9.6.2). The category of locally affine objects is called the category of <u>K/k-prevarieties</u>.

<u>Remark 9.6.5</u> The objects of $\underline{\mathrm{la}(A)}$ could equally have been defined as objects in $\underline{R\text{-Top}}$ which are locally isomorphic to affine objects. The description of the category $\underline{\mathrm{la}(A)}$ includes the definition of morphisms between locally affine objects: these must be local homomorphisms on each stalk. Indeed, the fact that each of the functors in the examples above is full (which is not entirely trivial) depends on this.

<u>Proposition 9.6.6</u>   *If* $V \subset A^n$ *is an affine variety, and* U *is an open subset of* V, *then* $(U, O_V|_U)$ *is a prevariety.*

<u>Proof</u>   Since U is open, we have $U = V \cap (A^n \setminus W)$ where $W \subset A^n$ is an affine variety.   If W is the set of common zeros of the polynomials $f_1, \ldots, f_r$ and W_i is the zero-set of f_i, then $W = \cap W_i$ and $U = \cup U_i$ where $U_i = V \cap (A^n \setminus W_i)$.   Now since $O_V|_{U_i} = (O_V|_U)|_{U_i}$ it suffices to show that $(U_i, O_V|_{U_i})$ is affine.

Let $Z_i = \{x \in A^{n+1} : f(x_1, \ldots, x_n) = 0$ for all $f \in I(V)$ and $f_i(x_1, \ldots, x_n)x_{n+1} - 1 = 0\}$, so that $Z \subset A^{n+1}$ is an affine variety.   The projection $(x_1, \ldots, x_{n+1}) \longmapsto (x_1, \ldots, x_n)$ from A^{n+1} to A^n is a regular map and takes Z_i homeomorphically on to U_i, inducing and isomorphism $(Z_i, O_{Z_i}) \cong (U_i, O_V|_{U_i})$.

<u>Example 9.6.7</u>   If $V = A^1$ and $U = A^1 \setminus \{0\}$, then the affine variety $Z = \{(x,y) \in A^2 : xy - 1 = 0\}$ is isomorphic to U.

Suppose that our functor $A : \underline{C} \to \underline{R\text{-}Top}_0$ also satisfies the condition

(9.6.8) *for any affine object* (X,F) *and* $U \subset X$ *open,* $(U, F|_U)$ *is locally affine.*

This condition is satisfied for $A = D^k$ or H by definition of the categories concerned, and it is also satisfied for prevarieties by 9.6.6. Under this condition, for any object (X,F) in $\underline{la(A)}$ and any U open in X we have $(U, F|_U) \in Ob(\underline{la(A)})$.   So if $Aff(X,F)$ is the collection of open subsets U of X such that $(U, F|_U)$ is affine, then $Aff(X,F)$ is a sub-basis for the topology on X, and so $X = \lim_{\to}\{U : U \in Aff(X,F)\}$ and and $(X,F) = \lim_{\to}\{(U, F|_U) : U \in Aff(X,F)\}$ the first limit taken in $\underline{Top}$, and the second in $\underline{R\text{-}Top}$.   Elements of $Aff(X,F)$ are called <u>affine pieces</u> of (X,F).

More generally, for any diagram $\{(U_i, F_i): i \in I\}$ of affine (or locally affine) objects such that the underlying maps of topological spaces are homeomorphisms with open image, then the object $(X,F) = \varinjlim (U_i, F_i)$ is locally affine, and has $X = \varinjlim U_i$ as underlying space. The category $\underline{la(A)}$ is thus the closure of the category of affine objects with respect to limits of such diagrams. Properties of the category $\underline{C}$ with regards limits are then inherited by $\underline{la(A)}$, for example products.

Proposition 9.6.9 *If $\underline{C}$ is a category with finite products and*

$A: \underline{C} \to \underline{R\text{-Top}}_0$ *is a functor satisfying* (9.6.2) *and* (9.6.8) *then* $\underline{la(A)}$ *has finite products.*

Proof Suppose $L = (X,F)$ and $M = (Y,G)$ are locally affine. Write

$L = \varinjlim X_i$ and $M = \varinjlim Y_j$ where $X_i = (X_i, F|_{X_i})$ and $Y_j = (Y_j, G|_{Y_j})$ are affine. The category of affine objects has products since it is equivalent to $\underline{C}$ under the functor A. Let $Z = \varinjlim (X_i \amalg Y_j)$ and $pr_L: Z \to L$, $pr_M: Z \to M$ the obvious projection maps. We shall show that Z is a product in $\underline{la(A)}$.

Suppose $f: W \to L$ and $g: W \to M$ are two maps in $\underline{la(A)}$. For each pair i,j let $W_{i,j} = f^{-1}(X_i) \cap g^{-1}(Y_j)$. Now since $\Phi(W_{i,j})$ is open in $\Phi(W)$, it is locally affine, and so we can write $W_{i,j} = \varinjlim \{W_{i,j,r}: r \in I_{i,j}\}$ where each $W_{i,j,r}$ is affine. The restrictions $f_{i,j,r}$ of f and g to $W_{i,j,r}$ induce a map from $W_{i,j,r}$ to $X_i \amalg Y_j$ is the product in the category of affine objects. Taking the limit of these maps over all i,j,r we obtain a map $h: W \to Z$ such that $pr_L h = f$ and $pr_M h = g$. Uniqueness is immediate. We are now able to deal with the separation axiom.

Definition 9.6.10 If $\underline{C}$ be a category with products, and $A: \underline{C} \to \underline{R\text{-Top}}_0$ is a functory satisfying (9.6.2) and (9.6.8) then the locally-affine object (X,F) is <u>separated</u> if the image of X in $\Phi[(X,F) \quad (X,F)]$ is closed.

In general, $\Phi[(X,F) \pi (X,F)]$ will not be the topological product $X \times X$.   Clearly a separated object in $\underline{la}(D^r)$ is a Hausdorff manifold.

9.7   VARIETIES

<u>Definition 9.7.1</u> A <u>K/k-variety</u> is a separated K/k-prevariety. A <u>morphism</u> between two varieties is a morphism in the category of prevarieties.

If (X,O_X) is a variety and Y is open or closed in X then $(Y,O_X|_Y)$ is a variety. If Y is closed it is called a subvariety. Since A^n is a variety, any affine variety is indeed a variety. The variety (X,O_X) is <u>irreducible</u> if X is irreducible as a topological space. In this case any two non-empty open sets have a non-empty intersection, so the limit $k(X) = \lim_{\rightarrow}\{O_X(U): U \in Ouv(X)\}$ exists. The ring $k(X)$ is a field, called the field of <u>rational functions</u> on X (cf. §6.2). Clearly $k(X)$ is isomorphic to the field of rational functions on any affine piece of X. The transcendence degree of $k(X)$ over k is finite, and is called the <u>dimension</u> of X.

Suppose now that (X,O_X) is a K-variety. The point $x \in X$ is <u>simple</u> if the dimension of the tangent space $T_x = m_x/m_x^2$ at x is equal to the dimension of X otherwise it is called <u>singular</u>. The set of singular points of an irreducible variety form a subvariety of codimension at least one. The point $x \in X$ is <u>normal</u> if $O_{X,x}$ is integrally closed. The variety X is <u>smooth</u> (<u>normal</u>) if each point is simple (normal).

An automorphism α of the field K induces an automorphism of the K-algebra $K[A^n]$ and a homeomorphism of the affine space A^n for each n. If $V \subset A^n$ is an affine variety, then so is $\alpha(V)$, being the zero set of the ideal $\alpha I(V)$. The homeomorphism $\alpha: V \rightarrow \alpha(V)$ induces an isomorphism $(V,O_V) \rightarrow (\alpha(V),O_{\alpha(V)})$ in <u>Z-Top</u>, but since this is not in general a

134

morphism in $\underline{K\text{-Top}}$, it is not an isomorphism of varieties. The varieties V

and $\alpha(V)$ are said to be $\underline{conjugate}$. For example if V consists of a single

point $x \in A^1$ for which $x \neq \alpha(x)$, then the stalks at $x, \alpha(x)$ are both K

and the map between stalks induced by α is α, which is not a K-algebra

homomorphism.

However, if α is the identity on some subfield $k \subset K$ and V is an

affine K/k-variety, then $\alpha: (V, O_V) \to \alpha(V), O_{\alpha(V)})$ is an isomorphism in

$\underline{k\text{-Top}}_0$ so V and $\alpha(V)$ are isomorphic as K/k-varieties.

9.8   PROJECTIVE VARIETIES

By analogy with the affine case (§9.1) a subset of P^n is closed if it is the

set of common zeros of a finite set of homogeneous polynomials with

coefficients in k.   The topology on P^n with these closed sets is the

$\underline{Zariski\text{-}topology}$, or the $\underline{k\text{-}topology}$.   For each i,   the affine piece

$A_i = \{(x_0: \ldots : x_n) \in P^n: x_i \neq 0\}$ is open, and the natural map $A^n \to A_i$

(see §6.7) is a homeomorphism with the k-topologies.   As before we shall

often identify A_i with A^n by means of this homeomorphism.   The family

$\{A_i: i = 0, \ldots, n\}$ is an open covering of P^n.   For any $X \subset P^n$, let

$X_i = X \cap A_i$ and $X_{i,j} = X \cap A_i \cap A_j$.   If U is open in P^n, then U_i is

open in $A_i = A^n$.

$\underline{Definition\ 9.8.1}$ A function $f: U \to K$ on the open subset U of P^n is

$\underline{regular}$ if its restriction to U_i is regular for each i.

If for each i we have a regular function $f_i: U_i \to K$ such that f_i

and f_j agree on $U_{i,j}$ then the function $f: U \to K$ which is f_i on U_i

is regular.   This idea of glueing together compatible functions is contained

in the sheaf condition.

If $O_{P^n}(U)$ is the ring of regular functions on U, then (P^n, O_{P^n}) is a variety, with the open sets A_i as affine open sets. More precisely P^n is the colimit of the diagram of affine varieties

$$\coprod_{i,j} A_{i,j} \rightrightarrows \coprod_{i} A_i$$

If $V \subset P^n$ is a closed subset, then each V_i and $V_{i,j}$ is an affine variety. If $O_V = O_{P^n}|_V$ then (V, O_V) is a variety, and is the colimit of the diagram of affine varieties

$$\coprod_{i,j} V_{i,j} \rightrightarrows \coprod_{i} V_i$$

Such a variety (V, O_V) is called a <u>projective variety</u>.

<u>Example 9.8.2</u> The definition of a map between varieties is, roughly speaking, that it should be expressible as polynomial maps in affine pieces. To illustrate this point, we show that the function f from the elliptic curve $V \subset P^2$ defined by the homogeneous polynomial $ZY^2 - X^3 - Z^3$ to the projective line P^1 which takes $(z:x:y)$ to $(z:x)$ if $z \neq 0$ but to $(0:1)$ if $z = 0$ (see §8.1) is a map of K-varieties. Let $U = \{(z:x:y) \in P^2 : x \neq 0\}$ and $U' = \{(z:x:y) \in P^2 : y^2 - z^2 \neq 0\}$. The sets U, U' are open in P^2 and their intersections with V cover V. On $U \cap V$ the map f takes $(z:x:y)$ to $(z:x)$, while on U' it takes $(z:x:y)$ to $(x^2 : y^2 - z^2)$ since for $(z:x:y) \in U \cap U' \cap V$ we have $(z:x) = (z(y^2 - z^2) : x(y^2 - z^2)) = (x^3 : x(y^2 - z^2))$ $= (x^2 : y^2 - z^2)$.

If $V \subset A^n = A_0$ is an affine variety, its closure $\overline{V}$ in P^n is a projective variety whose intersection with A_0 is V. The homogeneous polynomials defining $\overline{V}$ are the homogenisations of the polynomials defining V. For example in the case of curves (cf. §6.2, 6.7) if $f \in k[X,Y]$ is a polynomial defining the curve $V(f) \subset A^2$, then $V(f^H) \subset P^2$ is the closure of $V(f)$, and is open and dense in $V(f^H)$.

In order to deal with varieties in affine and projective space one makes the following definition.

Definition 9.8.3   A quasi-projective variety is an open subset of a projective variety.

9.9   BIRATIONAL EQUIVALENCE REVISITED

Suppose X,Y are two irreducible varieties.   Two maps $f_i: U_i \to Y$, where U_i is open and non-empty in X, are equivalent if $f_1 = f_2$ on $U_1 \cap U_2$. An equivalence class of such maps is called a rational map from X to Y. A rational function on X is a rational map from X to A^1.   If $f: U \to Y$ represents a rational map F from X to Y, then the closure $\overline{f(U)}$ of $f(U)$ in Y is independent of the choice of f, and if $\overline{f(U)} = Y$, then F is called dominating.

If F is a dominating rational map from X to Y and G is a rational map from Y to Z, then we can compose G with F (by taking the equivalence class of the composition of representative) and obtain a rational map from X to Z.   Thus we have a category Rat(K/k) whose objects are irreducible K/k-varieties and whose morphisms are dominating rational maps. An isomorphism in this category is called a birational correspondence.

If $F: X \to Y$ is a map in Rat(K/k) represented by $f: U \to Y$ where $U \subset X$ is open, then for each open set W in Y we have a homomorphism from $O_Y(W)$ to $O_U(f^{-1}W) = O_X(f^{-1}W)$ which in the limit induces a homomorphism $F^*: k(Y) \to k(X)$.

Theorem 9.9.1   *A map* $F: X \to Y$ *in* Rat(K/k) *is an isomorphism (i.e. a birational correspondence if and only if* $F^*: k(Y) \to k(X)$ *is an isomorphism of fields.*

9.10   GAGA

The word gaga is an anacronym[131] for géometrie algébrique et géometrie
analytique.   If for $i = 1,2$ we have functors $A_i : \underline{C} \to \underline{R}_i\text{Top}_0$ we say that
A_2 is <u>continuous with respect to</u> A_1 if the following condition is satisfied.
(9.10.1) for each map $f: c \to c'$ in $\underline{C}$ such that $\Phi A_1(f)$ is a homeomorphism
with open (closed) image, then $\Phi A_2(f)$ is a homeomorphism with open (closed)
image.

Suppose for $i = 1,2$ we have functors $A_1 : \underline{C}_i \to \underline{R}_i\text{-Top}_0$ each satisfying
(9.6.2) and (9.6.8) and that $B: \underline{C}_1 \to \underline{la(A)}_2 \subset \underline{R}_2\text{-Top}_0$ is continuous with
respect to A_1.   There is a natural extension of B to a functor from
$\underline{la(A)}_1$ to $\underline{la(A)}_2$ which preserves separated objects, namely if X is in
$\underline{la(A)}_1$ then put $B(X) = \lim_{\to} B(X_i)$ where $X = \lim_{\to} X_i$ with X_i affine for A_1.
Example 9.10.2   Consider the two functors $H: \mathcal{C} \to \underline{\mathcal{C}\text{Top}}_0$ and
$D^r: \underline{C} \to \underline{R} \to \underline{R\text{Top}}_0$.   The functor D^r is continuous with respect to H but
H is not continuous with respect to D^r.   The induced functor from $\underline{la(H)}$
to $\underline{la(D^r)}$ associates to a complex manifold its underlying D^r-manifold.
Example 9.10.3   If $V \to A^n$ is an irreducible affine K/k-variety it is a
fortiori also a K-variety.   The corresponding functor from $\underline{la(K/k)} \to \underline{la(K)}$
associates to each K/k-variety its 'underlying' K-variety.

If $V \subset \mathcal{C}^n$ is a smooth affine $\mathcal{C}$-variety, then as a subset of $\mathcal{C}^n$ with
its usual topology, V is closed.   If V_{an} denotes the set V with this
'strong' topology, then (V_{an}, H_V) is a complex manifold, where $H_V = H|_{V_{an}}$
is the sheaf of holomorphic functions on V.   The above construction thus
associates to any smooth complex variety (V, O_V) a complex manifold (V_{an}, H_V)
in a natural way.   If V is a subvariety of $P^n(\mathcal{C})$ then V_{an} is a complex
submanifold of $P^n(\mathcal{C})_{an}$.

9.11   CYCLES

Throughout this section we fix an algebraically closed field K and by variety we mean K-variety.

Let V be a smooth connected quasi-projective variety. A <u>cycle</u> on V is an element of the free abelian group on the irreducible subvarieties of V. This group $C^*(V)$ is graded by codimension, and we write $C^r(V)$ for the components of grade r, so that $C^0(V) = Z$ generated by V, $C^n(V)$ is the free abelian group on the points of V, while $C^i(V) = 0$ for $i > n$. The elements of $C^1(V)$ are called <u>divisors</u>.

Two subvarieties X, Y of V intersect properly[132] if for each component Z of $X \cap Y$ we have $\operatorname{codim} Z = \operatorname{codim} X + \operatorname{codim} Y$. It is possible to associate to each such component Z and integer $I(X,Y;Z)$ called the intersection number which measures the degree of multiplicity of Z in $X \cap Y$. The definition[133] of this integer is beyond the scope of these notes, and is in terms of homological algebra. However, if X and Y intersect properly then their product X, Y is defined to be $\Sigma I(X,Y;Z)Z$ where the sum is taken over the irreducible components of $X \cap Y$. This product extends linearly over $C^*(V)$ and as far as it is defined is commutative and associative with $V \in C^0(V)$ as identity.

In order to extend the domain of definition of this product, there are various equivalence relations on $C^*(V)$ which essentially allow us to move cycles to make sure they intersect properly. Two cycles X_1, X_2 on V are <u>algebraically equivalent</u> if there is a smooth connected quasi-projective variety T and a cycle Z on $V \times T$ such that

(i) for each $t \in T$, the cycles Z and $V \times t$ intersect properly

(ii) there are two points $t_1, t_2 \in T$ such that $Z.(V \times t_i) = X_i \times t_i$.

This relation between cycles is an equivalence relation[134]. If we insist that $T = P^1$ we also obtain an equivalence relation called <u>rational</u>

<u>equivalence</u>. The following is often called the Moving Lemma[135].

<u>Lemma 9.11.1</u> *For any two cycles* X,Y *on* V *there is a cycle* Y' *on* V

rationally equivalent to Y *such that* X *and* Y' *intersect properly. If*

X,Y,Z *are cycles such that* X *is rationally (algebraically) equivalent to*

Y, *and* X.Z *and* Y.Z *are defined, then* X.Z *is rationally (algebraically)*

equivalent to Y.Z.

<u>Example 9.11.2</u> Let K be an algebraically closed field, and put $V = P^n(K)$.

Any two hyperplanes in V are rationally equivalent. For suppose L_1, L_2

are homogeneous linear polynomials defining the hyperplanes

$H_i = \{(x) \in V: L_i(x) = 0\}$ for $i = 1, 2$. The cycle

$Z = \{(x,(a:b)) \in V \times P^1(K): aL_1(x) + bL_2(x) = 0\}$ intersects each $V \times t$

properly, while $H_1 = Z.(V \times (1:0))$ and $H_2 = Z.(V \times (0:1))$.

The set of equivalence classes under algebraic (rational) equivalence form

a graded ring denoted by $C^*_{alg}(V)$ $(C^*_{rat}(V))$. The ring $C^*_{rat}(V)$ is called

the <u>Chow ring</u> of V.

A third important relation is numerical equivalence. Let $\varepsilon: C^*(V) \to Z$

be the composite of the projection $C^*(V) \to C^n(V)$ with the augmentation

$C^n(V) \to Z$, $\Sigma n_i P_i \mapsto \Sigma n_i$. For any two cycles X,Y for which X Y is defined,

the integer $\langle X,Y \rangle = \varepsilon(X.Y)$ is called the <u>intersection number</u> of X and Y,

and counts the number of points of intersection of X with Y, counting

multiplicities.

<u>Example 9.11.3</u> If V is a smooth projective F_q^a/F_q-variety and $\pi: V \to V$

is the Frobenius map induced by $x \mapsto x^q$ on F_q^a, then for each $m \in N$ the

graph $\Gamma_m \subset V \times V$ of π^m is a subvariety which intersects properly the

diagonal $\Delta \subset V \times V$. Moreover, each point of the intersection has[136]

multiplicity 1 and so $\langle \Delta, \Gamma_m \rangle$ is precisely the number of fixed points of

140

Two cycles X,Y on V are <u>numerically equivalent</u> if $\langle X,Z\rangle = \langle Y,Z\rangle$ for all cycles Z for which the products are defined.   The set of equivalence classes form a graded ring $C^*_{num}(V)$ and the pairing $\langle -,-\rangle\colon C^*_{num}(V) \times C^*_{num}(V) \to Z$ is non-singular.   Clearly $C^*_{num}(V)$ is torsion-free.

Suppose $f\colon V \to W$ is a morphism between smooth connected quasi-projective varieties of dimensions n,m.   If X is an irreducible subvariety of V then the closure[137] Z of $f(Y)$ in W is also irreducible.   We define $f_*(Y)$ to be the cycle $d.Z$, where $d = \deg(K(Z)/K(Y))$ if $\dim Z = \dim Y$, and $d = 0$ otherwise.   The function f_* extends linearly to a homomorphism $f_*\colon C^*(V) \to C^*(W)$ increasing degree by $m-n$, which preserves rational, algebraic and numerical equivalence.

If X is an irreducible subvariety of W when its inverse image $f^{-1}V$ is a subvariety of V (if it is non-empty), but may not be irreducible, and the various equivalence relations are not in general preserved.   In order to overcome this difficulty, one defines an 'inverse image counting multiplicities' as follows.   Suppose that X is a cycle on W.   The intersection of $V \times X$ with the graph Γ_f of f in $V \times W$ projects under $pr_V\colon V \times W \to V, (v,w) \longmapsto v$ onto $f^{-1}X$.   Now by the Moving lemma 9.11.1, there is a cycle X' rationally equivalent to X which intersects Γ_f properly.   We define f^*X to be $pr_{V*}(X'.\Gamma_f)$.   This cycle is well-defined up to rational equivalence, and it is easy to see that $X \longmapsto f^*X$ preserves codimension, and the various equivalence relations.   Moreover, it is multiplicative, and is related to f^* by the projection formula $f_*(f^*X.Y) = X.f_*Y$.

<u>Example 9.11.3</u>   Let K be the algebraic closure of F_q, and $\pi\colon K \to K$ the Frobenius map $x \longmapsto x^q$.   If $V = P^n(K)$, then the $n - r$ dimensional subspace

$X = \{(x) \in V: x_0 = x_1 = \ldots = x_{r-1} = 0\}$ is isomorphic to P^{n-r}, and $\pi(X) = X$. The field of rational functions $K(P^{n-r})$ is isomorphic to the field $K(A^{n-r})$ of rational functions on affine space, which is $K(X_1, \ldots, X_{n-r})$. The K-algebra homomorphism induced by π from $K(P^{n-r})$ to $K(P^{n-r})$ takes, X_i to X_i^q, and so the degree of the corresponding field extension is q^{n-r}, that is to say $\pi_*(X) = q^{n-r}(X)$.

The graph of $\pi, \Gamma \subset V \times V$ has codimension n, and $V \times V$ has codimension r. The intersection $Z = \Gamma \cap (V \times V)$ is $\{(v, \pi(v)): \pi(v) \in V\}$ = $\{(v, \pi(v)): v \in V\}$ which has dimension n-r, so codimension n + r. Thus the cycles Γ and $V \times V$ intersect properly, so $\Gamma .(V \times V) = aZ$ for some integer a (the intersection number). The projection from $V \times V$ onto the first coordinate maps Z to H, and so $\pi^*(X) = aX$. From the projection formula, we see that $\pi_*(\pi*X.V) = X.\pi_*V$, so $\pi_*(aX) = q^n X$. But $a\pi_*(X) = aq^{n-r}X$, and so $a = q^r$, i.e. $\pi*(X) = q^r X$.

<u>Example 9.11.4</u> Suppose V is a smooth connected projective curve. The diagonal $\Delta \subset V \times V$ is a subvariety and $\Delta V.(V \times x) = x$ for each point $x \in V$, so any two points of V are algebraically equivalent, and hence numerical equivalence implies algebraic equivalence. The converse is also true but more difficult[138]. Let Div_0 be the subgroup of $C^1(V)$ of divisors algebraically equivalent to 0 so that $\mathrm{Div}_0 = \mathrm{Ker}(\varepsilon: C^1(V) \to Z)$.

By 6.7.2, the points of V correspond to non-trivial spots on K(V) which are trivial on K. The image of the homomorphism $K(V)^* \to \mathrm{Div}_0$, $f \mapsto \Sigma \, \mathrm{ord}_x f.x$ is the group D_1 of <u>principal divisors</u>, and the quotient $J(V) = \mathrm{Div}_0/D_1$ is called the <u>Jacobian</u> of V.

Suppose that K is the algebraic closure of the finite field k of order q, and $G = \mathrm{Gal}(K/k)$. If V is a K/k-variety, then the field k(V) is an A-field. If S is a non-trivial spot on k(V) then the set of

points of V over which S lies is (by 6.7.2) finite and constitutes an
orbit of G.   If d(S) is the sum of these points then (using the notation
of §5.4) we see that d induces a homomorphism from $d: Fr(P) \to C^1(V)$ such
that $\varepsilon d(S) = \deg(S)$.   So d maps G_0 into Div_0 and induces a
monomorphism from the divisor class group $D_0 = G_0/G_1$ of k(V) to the
Jacobian of V, i.e. to the divisor class group of K(V).   The group G
acts on C*(V), and hence on J(V).   Clearly D_0 the image is the subgroup
of J(V) fixed by π.   Now J(V) can be naturally given[139] the structure of
a smooth projective variety over k, so the subgroup D_0 is finite for
obvious geometric reasons.

9.12   THE WEIL CONJECTRUES

We have now developed enough language to be able to state the Weil
conjectures[140].

   Suppose V is a projective F_q^a/F_q-variety of dimension n, which as an
F_q^a-variety is smooth.   If $\pi: V \to V$ is the Frobenius map (induced by
$x \mapsto x^q$ on F_q^a), let ν_m be the number of fixed points of ν^m, so that
$\nu_m = |\{x \in V: \deg(x) \text{ divides } m\}|$ where deg(x) is the degree of the
residue field $0_{V,x}/m_x$ over F_q.   Put $Z_V(t) = \exp \Sigma \nu_m t^m/m$.   The
conjectures, all of which have now been proved[141], are as follows.

(9.12.1)   *$Z_V(t)$ is a rational function of* t.

(9.12.2)   *There is a functional equation*

$$Z_V(1/qt) = (qt^2)^{\chi/2} Z_V(t) \quad \text{for some } \chi \in \mathbb{Z}.$$

(9.12.3)   *There are polynomials* $P_i(t)$ *for* $i = 0,\ldots,2n$ *with* $P_i(0) = 1$
such that

$$Z_V(t) = \prod_i (P_i(t))^{(-1)^{i+1}}$$

(9.12.4)   *The roots of* $P_i(t)$ *have absolute value* $q^{-i/2}$.

This conjecture is called the <u>Riemann Hypothesis</u>.

In order to make precise the next conjecture, we really need the language of schemes. However, it is convenient and illustrative to present a naive version here. Suppose K is a field of characteristic O and S a spot on K with residue field $K(S) = F_q$, and that $f_1,\ldots,f_r$ are homogeneous polynomials with coefficients in R_S whose reductions mod m_S in $K(S) = F_q$ define the variety V, and that in addition for some embedding $\sigma:K \to C$ the complex projective variety X defined by the homogeneous polynomials $\sigma f_1,\ldots,\sigma f_r$ is smooth and has (complex) dimension n.

(9.12.5)   *The degree of the polynomial* P_i *is the i-th Betti number in* **(9.12.2)** *is the Euler-characteristic of the complex manifold* X_{an}.

Each of these conjectures depends on the previous one for its meaning. The next conjecture does not, and can be considered as a generalisation of the finiteness of ideal class group theorem (5.4.4).

(9.12.6)   *The group* $C^*_{num}(V)$ *of numerical equivalence classes is finitely generated.*

Along with these six precise conjectures, there is a 'conjecture' that there should be a 'cohomology theory' for varieties with which one could prove these conjectures as one can prove thoir complex analogues. In the next chapter, we indicate how these analogues are proved, except for the Riemann hypothesis,using cohomology. We note that since any two open sets of an irreducible variety intersect, the usual cohomology theories applied to a variety with the Zariski topology are all trivial. The Riemann Hypothesis is a much deeper conjecture, and we discuss complex analogues of this in Chapter 11.

10 Complex manifolds

We have seen in §9.10 that a smooth complex manifold $(V,0_V)$ naturally

determines a complex manifold (V_{an},H_V) of the same dimension.   For example,

if $V \subset C^m$ is a smooth affine variety of dimension n defined by polynomials

$f_1,\ldots,f_r \in C[Z_1,\ldots,Z_m]$ then for any point $v \in V$, since v is a simple

point, the Jacobian matrix $\|\frac{\partial f_i}{\partial z_j}\|$ has maximum rank at v, equal to n.

By the inverse function theorem,[142] there is an open set U in C^n

containing the origin O and a holomorphic function $h : U \to C^m$ which is a

homeomorphism onto its image, which is an open subset of V containing

$v = h(0)$.   Such functions h provide holomorphic charts on V_{an} in the

usual sense.

In order to distinguish between the two topologies on the underlying

point set of a smooth complex variety, we use the prefixes 'Z-' and 'S-'

in reference to the Zariski and strong (or usual) topologies respectively.

For example, since a Z-closed set is S-closed, the identity function

$1 : V_{an} \to V$ is continuous.   Now any regular function on V is S-continuous,

and indeed holomorphic, so there is a natural map $(V_{an},H_V) \to (V,0_V)$ in

$\underline{C\text{-Top}}_0$.   The following result is a generalisation of 7.3.1 and 7.4.3.

<u>Proposition 10.1.1</u> *The homomorphism* $0_{V,v} \to H_{V,v}$ *between the local rings*

at the point **v** *induced by the natural map* $(V_{an},H_v) \to (V,0_V)$ *becomes an*

isomorphism on completion.

This proposition will enable us to decide algebraically when a regular

map $f : V \to W$ between smooth complex varieties is a local homeomorphism in the S-topology.

<u>Definition 10.1.2</u>   A map $(f,\alpha) : (V,O_V) \to (W,O_W)$ between smooth complex varieties is <u>étale</u> at the point $v \in V$ if the map $O_{W,f(V)} \to O_{V,v}$ between stalks becomes an isomorphism on completion.   The map f is <u>étale</u> if it is étale at each point.

It is important to note that the concept of an étale map is independent of the S-topology, although it is clear from that $f : V \to W$ is étale at $v \in V$ if and only if $f : V_{an} \to W_{an}$ is a local homeomorphism (in the S-topology) at v .

<u>Example 10.1.3</u>   The regular map $z \to z^2$ from $C \setminus \{0\}$ to itself is an étale map, although it is not a local homeomorphism in the Z-topology.

This example indicates the value of étale maps: they would be local homeomorphisms if the topology were such that the inverse function theorem applied.

We shall be interested in projective varieties.   Let $\underline{V}_{sp}(\underline{C})$ be the category of smooth irreducible projective complex varieties and regular maps. For $V \in Ob \ \underline{V}_{sp}(\underline{C})$ the complex manifold V_{an} is S-connected, and S-compact, since it is an S-closed subspace of $P^n(C)_{an}$, which is S-compact.   A complex submanifold of $P^n(C)_{an}$ is called an <u>algebraic manifold</u>.   The following result is Chow's theorem[143].

<u>Theorem 10.1.4</u>   *If* (X,H_X) *is an algebraic manifold, then there is a smooth projective complex variety* (V,O_V) *such that* $(V_{an},H_V) \cong (X,H_X)$.

This leads to a generalisation of 7.5.4.

<u>Corollary 10.1.5</u>   *The functor from the category of smooth projective complex varieties and regular maps to the category of algebraic manifolds and holomorphic maps which takes* (V,O_V) *to* (V_{an},H_V) *is an equivalence of categories.*

146

In particular, any holomorphic map from V_{an} to W_{an} is regular, so
that any meromorphic function on V_{an} (being by definition a holomorphic
map from V_{an} to $S^2 = P^1(C)_{an}$) is rational. This explains why the study
of smooth projective complex varieties from the algebraic and analytic view-
points are equivalent. This is not true for smooth affine varieties as
example 10.1.3 shows.

10.2   COHOMOLOGY

We wish to apply cohomological methods to complex varieties in order to
prove analogous theorems to the Weil conjectures. We assume the reader to
be familiar with the basics of cohomology especially for orientable
manifolds[144], but for the sake of completeness we give a brief resumé here.

Firstly we remark that any two ordinary cohomology theories coincide on
complex varieties with the S-topology since[145] such spaces can be
triangulated[146]. We shall use theories of Čech type.

If X is a topological space, then the category $\underline{Ab(X)}$ of sheaves of
abelian groups on X is an abelian category[147] with sufficient injectives[148].
The functor $\Gamma : F \to F(X)$ from $\underline{Ab(X)}$ to $\underline{Ab}$, the category of abelian
groups, is left-exact[149]; its n-th derived functor is denoted by $H^n(X;-)$.
The group $H^n(X;F)$ is called the <u>n-th cohomology group with values in the
sheaf</u> F. This group can be calculated as the n-th homology group of the
chain complex obtained by applying the functor Γ to any injective
resolution of F, or more generally[150] to any resolution
$0 \to F \to F_0 \to F_1 \to \ldots$ of F by sheaves F_i for which $H^n(X;F_i) = 0$ for
$n > 0$. Examples of sheaves whose cohomology vanishes in positive
dimensions are injective, fine[151] or flabby[152] sheaves.

Any abelian group A determines a constant sheaf $\underline{A}$; we write
$H^n(X;A)$ for $H^n(X;\underline{A})$ and $H^n(X)$ for $H^n(X;Z)$.

There is a second approach which applies to constant sheaves. For any open cover $U = \{U_i : i \in I\}$ of X , let $N(U)$ be its nerve. This is the simplicial set whose n-simplices are those elements $(U_{i_0}, \ldots, U_{i_n})$ of U^{n+1} such that $U_{i_0} \cap \ldots \cap U_{i_n} \neq \phi$, with the obvious face and degeneracy maps

$$\partial_j (U_{i_0}, \ldots, U_{i_n}) = (U_{i_0}, \ldots, \hat{U}_{i_j}, \ldots, U_{i_n})$$

and

$$s_j (U_{i_0}, \ldots, U_{i_n}) = (U_{i_0}, \ldots, U_{i_j}, U_{i_j}, \ldots, U_{i_n}) .$$

If $V = \{V_j : j \in J\}$ is an open cover which refines U , then for each $j \in J$ we can choose an element $f(j) \in I$ such that $V_j \subset U_{f(j)}$. The function f induces a simplicial function from $N(V)$ to $N(U)$ taking $(V_{j_0}, \ldots, V_{j_n})$ to $(U_{f(j_0)}, \ldots, U_{f(j_n)})$. Although this simplicial map depends on the choice of f , its homotopy class does not[153]. If $H^n(X, U; A)$ denotes the n-th simplicial cohomology group of $N(U)$ with coefficients in the abelian group A , then the refinement V of U defines a 'restriction' homomorphism $H^n(X, U; A) \to H^n(X, V; A)$. The Čech cohomology group $\check{H}^n(X; A)$ is $\varinjlim \{H^n(X, U; A) : U \in \mathrm{Cov}(X)\}$ where $\mathrm{Cov}(X)$ is the lattice of open covers of X . When X is paracompact, the groups $H^n(X; A)$ and $\check{H}^n(X; A)$ are naturally isomorphic.

Alternatively we can consider the family $\{N(U) : U \in \mathrm{Cov}(X)\}$ as a pro-object[154] in the homotopy category <u>HSS</u> of simplicial sets, whose objects are simplicial sets and morphisms homotopy classes of simplicial maps, that is to say $\check{N}(X) = \{N(U) : U \in \mathrm{Cov}(X)\}$ is an sobject of <u>Pro-HSS</u>. This object is called the <u>Čech nerve</u> of X . The group $\check{H}^n(X; A)$ is the limit of the cohomology group $\{H^n(N(U); A) : U \in \mathrm{Cov}(X)\}$ in <u>Pro-Ab</u>.

A continuous map $f : X \to Y$ induces a homomorphism between cohomology groups which we denote by f^* or, if we wish to be precise about the cohomology theory and grading, $H^n(f)$.

When A is a ring, there is a natural product

$H^n(X;A) \times H^m(Y;A) \to H^{n+m}(X \times Y;A)$ which induces a multiplication map

$$H^n(X;A) \times H^m(X;A) \to H^{n1m}(X \times X;A) \xrightarrow{\Delta^*} H^{n+m}(X;A)$$

where $\Delta : X \to X \times X$ is the diagonal.   This multiplication gives $H^*(X;A)$

the structure of a graded-commutative graded A-algebra.

<u>Proposition 10.2.1</u>   *If V is a smooth connected projective complex variety*

of dimension n and A is a ring, then $H^r(V_{an};A)$ is a finitely generated

A-module, and is zero for r > 2n .

<u>Proof</u>   Trivial, since V_{an} can be triangulated using simplices of

dimension no greater that 2n .

<u>Notation 10.2.2</u>   For the rest of this chapter, V will denote a smooth

connected projective complex variety of dimension n ,   and we shall write

$H^*(W;A)$ for $H^*(W_{an};A)$ whenever W is a smooth complex variety.

By the Universal Coefficient theorem[155], there is a natural isomorphism

$H^r(V;K) \cong H^r(V) \otimes K$.   The complex manifold V_{an} is naturally oriented[156] as

a smooth real manifold of (real) dimension 2n ,   and so there is a canonical

generator $[V] \in H^{2n}(V)$ called the fundamental class of V ,   by means of

which we can identify $H^{2n}(V)$ with Z and $H^{2n}(V;K)$ with K .

For such a manifold, we have Poincaré duality[157]:

<u>10.2.3</u>   *multiplication* $H^r(V;K) \times H^{2n-r}(V;K) \to H^{2n}(V;K) = K$ *is non-singular.*

This implies that $H^r(V;K)$ and $H^{2n-r}(V;K)$ are naturally dual vector

spaces.   If $f : V \to W$ is a map in $V_{sp}(C)$ and dim W = n + d   then the

<u>umkehr homomorphism</u> $f_* : H^r(V;K) \to H^{r+2d}(W;K)$ is defined to be the adjoint

of the map $f^* : H^{2n-r}(W;K) \to H^{2n-r}(V;K)$.   This construction is functorial,

and f_*, f^* are related by the projection formula $f_*(f^*x.y) = x.f_*y$ for

$x \in H^*(W;K)$ and $y \in H^*(V;K)$.

<u>Remark 10.2.4</u>   We shall see in chapter 12 that if A is a finite abelian

group, then the cohomology group $H^n(V;A)$ has an algebraic description, independent of the S-topology.   In particular, this means that for any prime ℓ the group $H^n(V;Z_\ell) = \varprojlim_m H^n(V;Z/\ell^m Z)$ and the Q_ℓ vector space $H^n(X;Q_\ell) = Q_\ell \otimes H^n(X;Z_\ell)$ (which is finite dimensional by 10.2.1) are also algebraically determined.   In the following sections we shall work with cohomology groups $H^*(V;K)$ where K is a field of characteristic O,   and this remark shows that one such cohomology group is independent of the S-topology.

10.3   INTERSECTION THEORY

The reason that cohomological methods have any bearing on the problems is that intersection of subvarieties of V can be related to the product in the cohomology of V.

If X is a smooth irreducible subvariety of V of codimension m, then X_{an} is a connected complex submanifold of V_{an} and the element $cl_V(X) = i_*[X] \in H^{2m}(V;K)$,   where $i : X \to V$ is the inclusion map, is called the <u>fundamental class</u> of X.   More generally, each irreducible subvariety Y of V of codimension m determines[158] an element $cl_V(Y) \in H^{2m}(V;K)$ and the function cl_V extends linearly to a homomorphism $cl_V : C^*(V) \to H^{2*}(V;K)$ of graded abelian groups.   The homomorphism is natural[159] in the sense that for any map $f : V \to W$ in $\underline{V}_{sp}(\underline{C})$,   both the diagrams

<u>10.3.1</u>
$$
\begin{array}{ccc}
C^*(W) & \xrightarrow{f^*} & C^*(V) \\
\downarrow{cl_W} & & \downarrow{cl_V} \\
H^*(W;K) & \xrightarrow{f^*} & H^*(W;K)
\end{array}
\qquad
\begin{array}{ccc}
C^*(V) & \xrightarrow{f^*} & C^*(W) \\
\downarrow{cl_V} & & \downarrow{cl_W} \\
H^*(V;K) & \xrightarrow{f^*} & H^*(W;K)
\end{array}
$$

commute.   The importance of the class map is the following.

<u>Theorem 10.3.2</u>   *If* X,Y *are cycles on* V *which intersect properly, then*

$$\langle X,Y \rangle = cl_V(X_1)cl_V(X_2) \in K .$$

If $f : V \to V$ is a map in $\underline{V}_{sp}(\underline{C})$ with only finitely many fixed points[160], then the graph Γ_f of f is a subvariety of $V \times V$ which meets the diagonal $\Delta \subset V \times V$ properly. Now $\langle \Gamma_f , \Delta \rangle$ is by definition the number of fixed points of f counting multiplicities. By 10.3.3 this is equal to $cl_{V \times V}(\Gamma_f)cl_{V \times V}(\Delta) \in H^{2n}(V \times V;K)$. The formal analogy between algabraic equivalence and homotopy is reflected by the next result.

<u>Proposition 10.3.4</u> *If* X_1 , X_2 *are algebraically equivalent cycles on* V, *then* $cl_V(X_1) = cl_V(X_2)$.

<u>Proof</u> Suppose T is a smooth connected quasi-projective variety and Z is a cycle on $V \times T$ which intersects each $V \times t$ properly, and such that $Z.V \times t_i = X_i$ for points $t_i \in T$. We shall prove the proposition under the assumption that T is actually projective, the general case being similar, but needing[161] the theory of cohomology with compact supports and Poincaré duality for non-compact manifolds.

Consider the inclusion $i_t : V \to V \times T,\ v \mapsto (v,t)$. By naturality 10.3.1 we have $cl_V(X_i) = i_{t_i}^{*} cl_{V \times T}(Z)$, and the result follows since i_{t_1} and i_{t_2} are homotopic (T_{an} being path connected).

From 10.3.2 and 10.3.4 we see that the map cl_V induces a ring homomorphism from the ring $C^*_{alg}(V) \otimes K$ to $\underset{r}{\oplus}\, H^{2r}(V;K)$. This map is an isomorphism in some cases[162].

Now any Z-continuous map $f : V \to V$ induces a K-homomorphism $H^r(f)$ of the finite dimensional vector space $H^r(V;K)$, and so has a well-defined trace in K. The next result, the Lefschetz Trace theorem[163] is the key to the cohomological interpretation of the Z-function.

<u>Theorem 10.3.5</u> *If* $f : V \to V$ *is a regular map then*

$$cl_{V \times V}(\Gamma_f) . cl_{V \times V}(\Delta) = \sum_{r=0}^{2n} (-1)^r \, \mathrm{Tr}(H^r(f)) .$$

151

<u>Corollary 10.3.6</u> *The Euler characteristic of* V_{an} *is* $cl_{V \times V}(\Delta).cl_{V \times V}(\Delta)$.

<u>Corollary 10.3.7</u> *If* $f : V \to V$ *is a regular map* **with** *finitely many fixed points, then the number* $\langle \Gamma_f, \Delta \rangle$ *of fixed points counting multiplicities is*

$$\sum_{r=0}^{\infty} (-1)^r Tr(H^r(f)).$$

10.4 RATIONALITY

Suppose for the rest of this chapter that $\pi : V \to V$ is a regular isomorphism such that for each m, π^m has only a finite number ν_m of fixed points, each of multiplicity 1. By 10.3.7 we know that $\nu_m = \sum_{r=0}^{2n} (-1)^r Tr(H^r(\pi^m))$. Put $Z(t) = \exp \sum_{m=1}^{\infty} \nu_m t^m / m$. The following lemma is trivial.

<u>Lemma 10.4.1</u> *If* α *is an endomorphism of the finite dimensional vector space* E *over the field* F, *then*

$$\sum_{i=0}^{\infty} Tr(\alpha^i) t^i = \det(1 - t\alpha).$$

Now we have a theorem for $Z(t)$ analogous to 9.12.1 and 9.12.3.

<u>Theorem 10.4.2</u> *If* $\pi : V \to V$ *is a regular isomorphism in* $\underline{V}_{sp}(\underline{C})$ *such that for each* m, π^m *has only a finite number* ν_m *of fixed points each of multiplicity* 1, *then* $Z(t) = \exp \sum \nu_m t^m / m$ *is a rational function of* t *of the form*

$$\prod_{r=o}^{2n} P_r(t)^{(-1)^{r+1}}$$

where $P_r(t)$ *is an integral polynomial of degree* β_r, *the r-th Betti number of* V_{an}, *satisfying* $P_r(0) = 1$, *and with* $P_0(t) = 1 - t$ *and* $P_{2n}(t) = 1 - Dt$, *where* D *is the topological degree of* $\pi : V_{an} \to V_{an}$.

<u>Proof</u> This is obvious from 10.4.1 and the remarks preceding it, with $P_r(t) = \det(1 - tH^r(\pi))$, except possibly for the fact that $P_r(t)$ is integral. This is because $H^r(\pi) : H^r(V) \otimes K \to H^r(V) \otimes K$ is induced by a homomorphism $H^r(\pi) : H^r(V) \to H^r(V)$ of finitely generated abelian groups.

We note that the zeros and poles of $Z(t)$ are at reciprocals of

152

algebraic integers, and that the Riemann Hypothesis (9.12.4) is concerned
with the eigenvalues of $\pi^* : H^*(V;K) \to H^*(V;K)$.

10.5   THE FUNCTIONAL EQUATION

The functional equation is essentially a restatement of Poincaré duality.

<u>Theorem 10.5.1</u>   *Under the hypotheses of* 10.4.2 *we have*

(10.5.1.1) $\det(1 - tH^r(\pi)) = \det(1 - tDH^{2n-r}(\pi^{-1}))$

(10.5.1.2) $Z(1/Dt) = (D^2 t)^{\chi/2} Z(t)$, *where* χ *is the Euler characteristic of* V_{an}.

<u>Proof</u>   Replacing K by a larger field if necessary, we can assume that D has a $2n$-th root a in K. If we put $g_r = a^{-r}H^r(\pi)$, and $\pi_r = H^r(\pi)$, then $g_* : H^*(V;K) \to H^*(V;K)$ is a K-algebra homomorphism with $g_{2n} = 1$. The commutativity of the diagram

$$
\begin{array}{ccc}
H^r(V;K) \otimes H^{2n-r}(V;K) & \longrightarrow & H^{2n}(V;K) = K \\
{\scriptstyle g_r \otimes g_{2n-r}}\Big\downarrow & & {\scriptstyle g_{2n}}\Big\downarrow \quad {\scriptstyle 1}\Big\downarrow \\
H^r(V;K) \otimes H^{2n-r}(V;K) & \longrightarrow & H^{2n}(V;K) = K
\end{array}
$$

together with Poincaré duality shows that g_r and g_{2n-r} are adjoint maps and so $\det(1 - tg_r) = \det(1 - tg_{2n-r}^{-1})$. Replacing g_r with $a^{-r}H^r(\pi)$ we obtain equation (10.5.1.1). This equation implies the two equations

$$\det(\pi_{2n-r}) \det(1 - t\pi_r) = (-Dt)^{\beta_{2n-r}} \det(1 - (tD)^{-1}\pi_{2n-r})$$

$$\det(\pi_{2n-r}) \det(\pi_r) = D^{\beta_r} = D^{\beta_{n-r}}$$

where $\beta_r = \dim_K H^r(V;K)$ is the r-th Betti number of V_{an}, and these imply that

$$\left| \prod_{r=0}^{2n} \det(\pi_{2n-r})^{(-1)^{r+1}} \right| Z(t) = (-Dt)^{-\Sigma(-1)^r \beta_{2n-r}} Z(\tfrac{1}{Dt})$$

and

$$\left| \prod_{r=0}^{2n} \det(\pi_{2n-r})^{(-1)^{r+1}} \right| = D^{-\frac{\chi}{2}}$$

and (10.5.1.2) follows immediately.

When V is a curve we have $\chi = 2 - 2g$ where g is the topological genus of V_{an}, or by 7.6.2, the algebraic genus of the field $M(V_{an})$ or meromorphic functions on the Riemann surface V_{an}. The functional equation is now $Z(\frac{1}{Dt}) = (D^2 t)^{1-g} Z(t)$ which should be compared with equations 3.3.5 and 5.5.3.

10.6   REDUCTION

In this section we give some of the evidence in support of the Weil conjecture (9.12.5) concerning reduction.   First if
$f(Z,X,Y) = ZY^2 - a_3 X^3 - a_2 X^2 Z - a_1 XZ^2 - a_0 Z^3$ is an integral polynomial defining a smooth elliptic curve V_C in $P^2(C)$, then the non-zero Betti numbers of the associated Riemann surface are $1, 2, 1$ in dimensions $0, 1$ and 2.   If p is a prime other than $2, 3$ such that $p \nmid a_3$ and the polynomial $a_3 X^3 + a_2 X^2 + a_1 X + a_0$ has distinct roots over F_p (almost all primes p have these properties) then the reduction of $f \bmod p$ defines a smooth projective curve V_p over F_p.   This curve also has genus 1 by the definition in §3.3 and 8.3.1 confirms 9.12.5 in this case.

Let k be the algebraic closure of the field F_p of order p and let k_m be the unique subfield of k of order p^m.   The projective space $P^n(k)$ can itself be considered as the reduction $\bmod p$ of complex projective space $P^n(C)$.   The set of point in $P^n(k)$ fixed by π^m is $P^n(k_m)$ which has $\nu_m = (p^{mn}-1)/(p^m-1)$ points, and so
$Z(t) = \exp \Sigma \, (p^{mn} - 1) t^m / m (p^m - 1) = \prod_{r=0}^{n} (1 - p^r t)^{-1}$.   This again confirms 9.12.5, since the r-th Betti number of $P^n(C)_{an}$ is 0 if r is odd or greater than $2n$, but 1 otherwise.

Our third example is of the Grassmanian   For any field F the

Grassmanian $G_{r,s}(F)$ is the set of r-dimensional subspaces of the vector space $F^{r,s}$. The group $GL_{r+s}(F)$ acts transitively on $G_{r,s}(F)$ in the obvious way, and the stabilizer of $F^r \oplus 0$ in F^{r+s} is the subgroup $H_{r,s}(F)$ of matrices of the form $\begin{pmatrix} A & C \\ O & B \end{pmatrix}$ where $A \in GL_r(F)$ and $B \in GL_s(F)$, and so $G_{r,s}(F) = GL_{r+s}(F)/H_{r,s}(F)$.

For $E \in G_{r,s}(F)$, the r-th exterior power $\lambda^r E$ is a 1-dimensional subspace of F^N where $N = \binom{r+s}{r}$ The function $E \mapsto \lambda^r E$ from $G_{r,s}(F)$ to $P^{N-1}(F)$ is injective and its image is[164] the set of common zeros of a family of homogeneous polynomials with integral coefficients. This implies that the Grassmanian $G_{r,s}(k)$ is the reduction mod p of the complex Grassmanian $G_{r,s}(C)$. Now the set of fixed points of π^m on $G_{r,s}(k)$ is precisely $G_{r,s}(k_m)$. Since the order of $GL_r(F)$ when F is a field of order q is $\prod_{i=0}^{r-1}(q^r - q^i) = a_r(q)$ say, the order of $G_{r,s}(F)$ is

$$\frac{a_{r+s}(q)}{a_r(q) a_s(q) . q^{rs}} = f_{r,s}(q) = f(q) \quad \text{say.}$$

It follows that the Z-function for the variety $G_{r,s}(k)$ is $\exp \Sigma f(p^m) t^m/m$.

Now $f(X)$ is[165] an integral polynomial in X, say $\Sigma b_i X^i$ in fact with non-negative coefficients. It is easy to see that

$$Z(t) = \exp \sum_{m=1}^{\infty} \sum_i b_i p^{mi} t^m/m = \prod(1 - p^i t)^{-b_i}.$$

It is well-known[166] that the Betti numbers of $G_{r,s}(C)_{an}$ vanish in odd dimensions, which is reflected by the fact that $Z(t)$ has no zeros. But more important is the observation that the coefficient b_i of X^i in $f(X)$ is[167] precisely β_{2i} as 9.12.5 suggests is should be.

Before this particular Weil conjecture was proved, many more examples were found[168] which confirmed this conjecture.

Suppose $X \in C^r(V)$ and $cl_V(X) = 0 \in H^{2r}(V;K)$. For any cycle Y

$cl_V(X.Y) = cl_V(X)cl_V(Y) = 0$, so $X.Y = 0$ since the class map

$cl_V : C^n(V) \to H^{2n}(V;K)$ is injective, and so X is numerically equivalent to

0.   So if I^r is the image of $C^r(V)$ in $H^{2r}(V;K)$ under the map cl_V,

there is a natural map from I^r onto $C^r_{num}(V)$, and hence from $K \otimes I^r$

onto $K \otimes C^r_{num}(V)$.   Now $K \otimes I^r$ is the K-subspace of $H^{2r}(V;K)$ generated

by I^r, and so is finite dimensional by 10.2.1, and so $K \otimes C^r_{num}(V)$ is also

finite dimensional.   Recall that the pairing $C^r_{num}(V) \otimes C^{2n-r}_{num}(V) \to Z$ is

non-singular.

<u>Lemma 10.7.1</u> *If* A,B *are abelian groups and* $m : A \otimes B \to Z$ *is a non-*

singular pairing such that for some field K *of characteristic* 0 *the*

K *vector space* $K \otimes B$ *is finite dimensional, then* A *and* B *are finitely*

generated.

<u>Proof</u>   Choose a finite set $b_1,...,b_n$ of elements of B which form a basis

for $K \otimes B$, and let $C \subset B$ be the subgroup they generate.   From the

commutative diagram

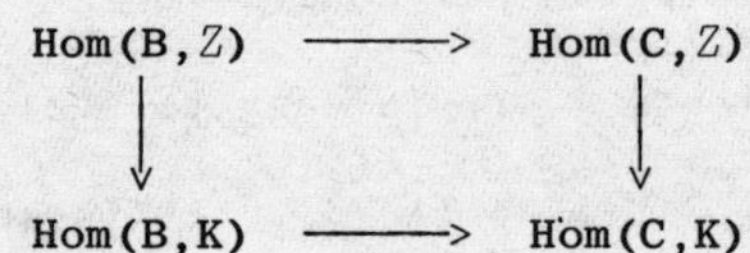

we see that $Hom(B,Z) \to Hom(C,Z)$ is injective.   The group $Hom(C,Z)$ is

finitely generated since C is, and so $Hom(B,Z)$ is finitely generated.

The non-singularity of the pairing implies that the adjoint map $A \to Hom(B,Z)$

is injective, and so A is finitely generated.   The same proof now shows

that B is finitely generated.

    Applying this lemma to the case of numerical equivalence classes, we

obtain^{169} the following.

<u>**Theorem 10.7.2**</u>   *For* **V** *a smooth connected projective complex variety, the group* $C^*_{num}(V)$ *is finitely generated.*

11 Hodge theory

11.1   SHEAVES OF MODULES

Since the category $\underline{PreSh(X)}$ of presheaves on X is a category of $\underline{Set}$-valued functors, any construction one can make functorially in Sets can be made 'pointwise' in the category $\underline{PreSh(X)}$. If these constructions involve only finite diagrams then the analogous construction applied to sheaves will again yield sheaves. For example, the category $\underline{Ab(X)}$ is the category of abelian groups in $\underline{Sh(X)}$, and (X,R_X) is a ringed-space if and only if R_X is a ring in $\underline{Sh(X)}$.

A continuous map $f : X \to Y$ induces functors $f_* : \underline{Sh(X)} \to \underline{Sh(Y)}$ and $f^* : \underline{Sh(Y)} \to \underline{Sh(X)}$ which are [170] adjoint: $\mathrm{Hom}_{\underline{Sh(X)}}(f^*F,E) \cong \mathrm{Hom}_{\underline{Sh(Y)}}(F,f_*E)$. Hence the functor f_* preserves limits, and f^* colimits. Suppose $a = (f,\alpha) : (X,R) \to (Y,S)$ is a morphism of ringed spaces. If   an R-module, then f_*M is an f_*R module, and hence an S-module via the ring homomorphism $\alpha : S \to f_*R$ (restriction of scalars). This S-module is denoted by a_*M. Conversely if N is an S-module, then f^*N is an f^*S module. Now the map $\alpha : S \to f_*R$ induces $f^*(\alpha) : f^*S \to f^*f_*R$ and if we compose this with the adjoint map $f^*f_*R \to R$ (corresponding to $1 \in \mathrm{Hom}(f_*R,f_*R)$) we have a ring homomorphism $f^*S \to R$. The R-module $R \otimes_{f^*S} f^*N$ is denoted by a^*M. The functors $a_* : \underline{R\text{-Mod}} \to \underline{S\text{-Mod}}$ and $a^* : \underline{S\text{-mod}} \to \underline{R\text{-Mod}}$ are again adjoint

An R-module which is isomorphic to the R-module R^n is said to be $\underline{free}$ $\underline{of\ rank}$ n, and a module M which has the property that there is an open covering $\{U_i : I \in I\}$ of X such that $M|_{U_i}$ is a free $R|_{U_i}$-module of rank n is called <u>locally-free of rank</u> n .   Clearly the direct sum if M,N are free (locally-free) R-modules then so are $M \oplus N$, $M \otimes N$ $Hom_R(M,N)$ and $\lambda^i M$. The group Pic(R) of isomorphism classes of locally-free R-modules of rank 1 under tensor product is called the <u>Picard group</u> of R   (cf. §5.4) If $a : (X,R) \to (Y,S)$ is a map of ringed spaces and N is a free (locally-free) S-module, then a^*N is a free (locally-free) R-module.   In general a_* does not preserve freeness.

Let (M,D_M) be a smooth manifold, and let $D_{M,C}$ be the complexification $C \otimes_R D_M$ of M , i.e. the sheaf of germs of smooth complex valued functions on M .   The ring homomorphisms $i : R \to C$ and $c : C \to C$, $z \to \bar{z}$ induce by composition morphisms $i : (M,D_{M,C}) \to (M,D_M)$ and $c : (M,D_{M,C}) \to (M,D_{M,C})$ of ringed spaces which induce functors between the associated categories of modules.

<u>Lemma 11.1.1</u> *If E is a free (locally-free) $D_{M,C}$-module of rank n then i^*E is a free (locally-free) D_M-module of rank 2n ,   and there are natural equivalences $i_* i^* = 1 \oplus 1$, $i^* i_* = 1 \oplus c^*$ where 1 is the identity functor on the relevant category.*

<u>Proof</u>   It suffices to prove the result when M is a point, when the proof is trivial.

<u>Definition 11.1.2</u>   If E is a $D_{M,C}$-module, then the $D_{M,C}$-module c^*E is denoted by $\bar{E}$ and called the <u>conjugate</u> module.

If $(X,R) \in$ Ob <u>Z-Top</u>$_0$ (that is each stalk R_x is a local ring) then a map $f : M \to N$ of R-modules is called <u>local</u> if $f_x(m_x M_x) \subset m_x N_x$.   The sub-category of <u>R-mod</u> of modules and local maps is denote by <u>R-Mod</u>$_0$.

<u>Example 11.1.3</u>   If $f : E \to X$ is a continuous family of real vector spaces over X ,   then the sheaf of sections $\tilde{E}$ is a module over the sheaf R of germs of real valued functions on X .   The map $f : E \to X$ is locally-trivial

(i.e. E is a vector bundle) if and only if $\tilde{E}$ is locally-free. The fibre $E_x = p^{-1}(x)$ of the vector bundle E is naturally isomorphic to the $R = R_x/m_x$ module $E_x/m_x E_x$. A vector bundle map induces a local map between the corresponding sheaves, and[171] the category of vector bundles and vector bundle maps is equivalent to $\underline{R\text{-Mod}}_0$. The condition that maps of sheaves be local maps corresponds to vector bundle maps being fibre preserving.

By virtue of the previous example and its generalisations, the concepts of vector bundles and locally free sheaves are interchangeable. We choose to work with sheaves. If (X, O_X) is a smooth complex variety, then a locally free sheaf over $(X, O_X), (X_{an}, H_X), (X_{an}, D_X), (X_{an}, D_{X,C}), (X_{an}, C_X)$ or (X_{an}, R_X) is (the sheaf of sections of) an algebraic, holomorphic, smooth, complex smooth, complex or real vector bundle respectively. The various natural maps between these ringed spaces induce transformations between the vector bundles of the corresponding types.

11.2   TANGENTS

In this and the next section we describe simultaneously in the language of sheaves the tangent and cotangent bundles of a smooth K-variety, real manifold and complex manifold. We start with the algebraic analogue of differentiation.

<u>Definition 11.2.1</u>   If A is an R-algebra and B an A-module, then an R-<u>derivation</u> from A to B is an R-module homomorphism $d : A \to B$ such that $d(a_1 a_2) = a_1 d(a_2) + a_2 d(a_1)$.

The set $\mathrm{Der}_R(A, B)$ of R-derivations from A to B is an A-module. If $d : A \to B$ is an R-derivation then $d(1) = d(1.1) = 2d(1)$, so $d(1) = 0$ and d vanishes on R. Inductively we see that $d(a^n) = na^{n-1}d(a)$ and so if f

is a polynomial in a over R then $d(f(a)) = f'(a)d(a)$. If
$d_1,d_2 \in \mathrm{Der}_R(A,A)$ then the map $[d_1,d_2] : a \mapsto d_1d_2(a) - d_2d_1(a)$ is also an
R-derivation and this bracket operation gives $\mathrm{Der}_R(A,A)$ the structure of a
Lie algebra over R .

<u>Example 11.2.2</u> If A is the *R*-algebra of smooth functions on an open
subset U of R^n then $f \mapsto \partial f/\partial X_i$ is an *R*-derivation from A to A .

<u>Example 11.2.3</u> If A is an R-algebra augmented over the ring R by
$e : A \to R$ and $m = \mathrm{Ker}(e)$, then the R-module homomorphism $a \mapsto a - e(a)$
from A to m induces an R-derivation from A to m/m^2 . In particular
if $R = R$ and A is the algebra of the previous example augmented over R
by $f \mapsto f(x)$ for some $x \in U$, then m is the ideal of functions which
vanish at x and the derivation $A \to m/m^2$ associated to f its derivative
at x .

Next we examine derivations of fields.

<u>Proposition 11.2.4</u> *If K/k is an extension and E/K is a finite separable
extension of fields, then any k-derivation from K to K extends uniquely
to a k-derivation from E to E , i.e. $\mathrm{Der}_k(K,K) \otimes_K E = \mathrm{Der}_k(E,E)$.*

<u>Proof</u> Let e be a primitive element of E and $f(X) = X^n + \sum_{i=0}^{n-1} a_i X^i$ its
minimum polynomial, so that $f(e) = 0$. If d is a derivation from E to
E then $d(f(e)) = 0$ implies that $f'(e)d(e) + \sum d(a_i)e^i = 0$ so d(e), and
hence d, is determined by the restriction of d to K .

To prove existence, suppose $d : K \to K$ is a k-derivation. The field E
is a $K[X]$ module via the K-algebra homomorphism $K[X] \to E$
taking X to e . Let $d : K[X] \to E$ be the k-homomorphism

$$d(\Sigma\, b_j X^j) = \Sigma\, (d(b_j)e^j + b_j c^j)$$

where $c = -\sum_{i=0}^{n-1} d(a_i)e^i/f'(e) \in E$. The map d is clearly a k-derivation.
Now $d(gf) = g(e)df + f(e)dg = 0$ since $df = 0$ by construction, and so d

161

factors through $E = K[X]/\langle f \rangle$ as a k-derivation from E to E which agree with d on K.

Using this result, we have a link between derivations and separability.

<u>Proposition 11.2.5</u>   *If E/k is a finitely generated extension of transcendence degree r, then the E-vector space $\mathrm{Der}_k(E,E)$ has dimension r.*

<u>Proof</u>   Since E is a finite separable extension of $K = k(X_1,\ldots,X_r)$ it suffices by 11.2.4 to prove the result when $E = K$.   For each i we have a derivation $d_i : f \mapsto \partial f/\partial X_i$.   If d is any k-derivation then $d' = \Sigma\, d(X_i)d_i$ agrees with d on X_i for each i and hence on the whole of K.   Clearly the elements $d_1,\ldots,d_r$ are linearly independent over K.

There is a converse[172] to this result: if K/k is an extension such that $\mathrm{Der}_k(K,K) = 0$ then K/k is a separable extension.   For example, if K is a field of characteristic O then $\mathrm{Der}_Q(K,K) = 0$ if and only if K/Q is an algebraic extension.

If (X,R) is a ringed space, A is an R-algebra and B is an A-module, then $\mathrm{Der}_R(A,B)$ is defined to be the sheaf of (sheaf) derivations from A to B .   If $U \subset X$ is open, then the natural map $\mathrm{Der}_R(A,B)(U) \to \mathrm{Der}_{R(U)}(A(U),B(U))$ (which is not in general an isomorphism) induces in the limit a map from the stalk of $\mathrm{Der}_R(A,B)$ at x to $\mathrm{Der}_{R_x}(A_x,B_x)$.   The sheaf $T_R = \mathrm{Der}_R(R,R)$ is called the <u>tangent sheaf</u> of R, and is an R Lie algebra.   Clearly if U is open in X then the tangent sheaf $T_{R|_U}$ is naturally isomorphic to $T_R|_U$ as $R|_U$ modules.   If R_x is a local ring for each $x \in X$, then the R_x/m_x vector space $T_x/m_x T_x$ is called the <u>tangent space</u> at x .

<u>Example 11.2.6</u>   If (X,D_X) is a smooth connected (real) manifold of dimension n with tangent sheaf $T_X = \mathrm{Der}_{R_X}(D_X,D_X)$ then T_X is precisely the sheaf of differentiable sections of the tangent bundle, so for U open

162

in X, the elements of $T_X(U)$ are the <u>smooth vector fields</u> on U. In particular, T_X is a locally-free D_X-module of rank n. In this case[173], the natural map $T_X(U) \to \mathrm{Der}_{D_X(U)}(D_X(U),D_X(U))$ is an isomorphism because of the existence of smooth partitions of unity.

<u>Example 11.2.7</u>   If (X,H_X) is an n-dimensional complex manifold with tangent sheaf T_{H_X}, then for U open in X, $T_{H_X}(U)$ is the set of <u>holomorphic vector fields</u> on U, and so T_{H_X} is a locally-free H_X module of rank n.

11.3   COTANGENTS

Suppose R is a ring and A is an R-algebra. Let I be the kernel of the multiplication map $A \otimes_R A \to A$, and let $p_i : A \to A \otimes_R A$ be the maps $p_1(a) = a \otimes 1$, $p_2(a) = 1 \otimes a$, and let d be the R-module homomorphism $p_1 - p_2$. The ideal I is generated by the image of d, and the quotient ring $A \otimes_R A/I^2$, called the ring of <u>principal parts of order 1</u>, is augmented over A by the multiplication map. The kernel of the augmentation is the ideal I/I^2 which can be considered as an A-module via either of the maps $p_1,p_2 : A \to A \otimes_R A/I^2$ since for any $x \in I$ and $a \in A$, $p_1(a)x - p_2(a)x = d(a)x \in I^2$. The A-module I/I^2 is denoted by $\Omega_{A/R}$ and is called the module of <u>1-differentials</u>. The function $d : A \to \Omega_{A/R}$ is a derivation, called the <u>exterior derivative</u>. If B is an A-module and $f : \Omega_{A/R} \to B$ is an R-module homomorphism, then $fd : A \to B$ is an R-derivation, and one can show[174]:

<u>Lemma 11.3.1</u>   *If A is an R-algebra and B an A-module, then the natural map $\mathrm{Hom}_R(\Omega_{A/R},B) \to \mathrm{Der}_R(A,B)$ is an isomorphism of A-modules. If A' is a second R-algebra, then the natural map $\Omega_{A/R} \otimes_A A' \to \Omega_{A \otimes_R A'/A'}$ is an isomorphism.*

In particular, we see that $\mathrm{Der}_R(A,A)$ is the dual of the A-module

163

$\Omega_{A/R}$. Using this notation, we see that if k is a field of characteristic 0, then K is algebraic if and only if $\Omega_{K/Q} = 0$.

Now let (X,O_X) denote either a connected smooth K-variety, smooth real manifold or complex manifold of dimension n (in the appropriate sense) and let $(X \amalg X, O_{X \amalg X})$ be the product $(X,O_X) \amalg (X,O_X)$ in the corresponding category of locally-affine objects, with p_1, p_2 the projection maps, and Δ the diagonal. If I' is the ideal in $O_{X \amalg X}$ of functions vanishing on the closed subspace $\Delta(X)$, let I be the ideal $\Delta^* I'$ of the ring $B = \Delta^* O_{X \amalg X}$ of sheaves on X. The ring B/I^2 is augmented over O_X and is called the sheaf of principal parts of order 1. The kernel of the augmentation is I/I^2 which is an O_X-module via either of the maps p_1 or p_2, and this O_X-module denoted by Ω_X is the sheaf of 1-differentials on X. Clearly for U open in X, then the $O_U = O_X|_U$ module $\Omega_X|_U$ is naturally isomorphic to Ω_U, and so the smoothness of X implies that Ω_X is locally-free, of rank n.

The O_X-module homomorphism $d = p_1 - p_2 : O_X \to \Omega_X$ is called the exterior derivative, and as in 11.3.1, the natural map $\mathrm{Hom}_{O_X}(\Omega_X, B) \to \mathrm{Der}_{O_X}(O_X, B)$, where B is an O_X-module, is an isomorphism. In particular, the tangent sheaf T_X is the dual O_X-module to Ω_X. For any point $x \in X$ we have an isomorphism $\mathrm{Hom}_{O_{X,x}}(\Omega_{X,x}, m_x/m_x^2) \to \mathrm{Der}_{O_{X,x}}(O_{X,x}, m_x/m_x^2)$ and the derivation $O_{X,x}$ derivation from $O_{X,x} \to m_x/m_x^2$ described in 11.2.3 corresponds to an $O_{X,x}$-module homomorphism $\Omega_{X,x} \to m_x/m_x^2$ which induces an isomorphism $\Omega_{X,x}/m_x \Omega_{X,x} \to m_x/m_x^2$ of $O_{X,x}/m_x$ vector spaces, so that indeed the vector bundle corresponding to Ω_X has fibre at x the cotangent space (cf 9.4.2).

Example 11.3.2 If S^1 is the circle $\{(x,y) \in R^2 : x^2 + y^2 = 1\}$ then the coordinate maps $x,y : S^1 \to R$ are smooth. Applying the exterior derivative d to the element $x^2 + y^2 - 1$ we see that $x\,dx + y\,dy = 0 \in \Omega_X$. Now S^1

is covered by the two open sets $U = \{(x,y) : x \neq 0\}$ and $V = \{(x,y) : y \neq 0\}$. The elements $x\,dx \in \Omega_X(U)$ and $-y\,dy \in \Omega_X(V)$ restrict to the same element of $\Omega_X(U \cap V)$ and so since Ω_X is a sheaf define an element of $\Omega_X(X)$.

The r-th exterior power Ω_X^r of the O_X-module Ω_X is called the sheaf of __r-differentials__, the elements of $\Omega_X^r(U)$ being called __r-forms__ on U. This is a locally-free sheaf of rank $\binom{n}{r}$ and so is zero for $r > n$, and there is a natural isomorphism $\Omega_X^0 = O_X$. The sheaf $K = K_X = \Omega_X^n$, called the __canonical sheaf__, is locally free of rank 1. The direct sum $\oplus_r \Omega_X^r$ is the O_X-__exterior algebra of differentials__, which is graded-commutative, the product being denoted by $(a,b) \mapsto a \wedge b$. The following is trivial[175].

__Lemma 11.3.3__ *There is a unique O_X-module homomorphism $d : \oplus \Omega_X^r \to \oplus \Omega_X^r$ of degree 1 satisfying*

(11.3.3.1) $\quad d^2 = 0$

(11.3.3.2) $\quad d : O_X \to \Omega_X$ *is the exterior derivative*

(11.3.3.3) $\quad d$ *is an algebra homomorphism, i.e. for $a \in \Omega_X^r(U)$ and $b \in \Omega_X^s(U)$ with $U \subset X$ open, we have $d(a \wedge b) = da \wedge b + (-1)^r a \wedge db$.*

11.4 DIVISORS

Let (X, O_X) be a smooth connected projective K-variety. Suppose $i : Y \to X$ is the inclusion of an irreducible subvariety of codimension 1. The kernel I_Y of the map $O_X \to i_* O_Y$ is the sheaf of ideals of functions vanishing on Y, and is a locally-free O_X-module of rank 1. For if $x \in Y$, then there is a connected open set U containing x such that $Y \cap U$ is the zero set of a single regular function on U (i.e. element of $O_X(U)$). The ideal $I_Y(U)$ of $O_X(U)$ is thus principal, and since $O_X(U)$ has no zero-divisors, $I_Y(U)$ is a free $O_X(U)$ module of rank 1. If $x \notin Y$, then x lies in the open set $U = X \setminus Y$, and $I_Y(U) = O_X(U)$ since $i_* O_Y(U) = 0$.

The function $Y \mapsto I_Y \in \text{Pic}(O_X)$ extends to a unique surjective homomorphism $D \mapsto I_D$ on the group $C^1(X)$ of divisors on X. Two divisors which have the same image are called <u>linearly equivalent</u>. One can show[176] that each element f of the field $K(X)$ of rational functions on X determines a divisor $(f) \in C^1(X)$ (cf §7.7, §5.4) and that the function $f \mapsto (f)$ is a homomorphism from $K(X)^*$ to $C^1(X)$ such that the sequence $0 \to K^* \to K(X)^* \to C^1(X) \to \text{Pic}(O_X) \to 0$ is exact (cf §5.4).

If $K = \mathcal{C}$, then each O_X-module M determines a $\mathcal{C}_X$-module $\mathcal{C}_X \otimes_{O_X} M$ and this defines a homomorphism from $\text{Pic}(O_X)$ to $\text{Pic}(\mathcal{C}_X)$, the group of (isomorphism classes) of 1-dimensional (continuous) complex vector bundles on X. This latter group is[177] isomorphic to $H^2(X;\mathbb{Z})$, the isomorphism given by the Chern class c_1. Moreover, the function $D \to c_1(\mathcal{C}_X \otimes_{O_X} I_D)$ from $C^1(X)$ to $H^2(X;\mathbb{Z})$ is precisely the class map. This suggests that if one had a cohomology theory for smooth K-varieties, then a first attempt at producing a class map would be to construct a theory of Chern classes[178].

11.5   REAL HODGE THEORY

Let (X,D_X) be a smooth real manifold, compact and connected of dimension n. In this section all sheaves will be understood as D_X-modules. Let T be the tangent sheaf and A^* the algebra of differential forms. If R_d denotes the constant sheaf at R with the discrete topology (i.e. the sheaf of germs of locally-constant functions) then there is a natural inclusion $i : R_d \to A^0 = D_X$. We have the <u>Poincaré lemma</u>[179].

<u>Theorem 11.5.1</u> *The sequence of sheaves of real vector spaces*

$$0 \to R_d \xrightarrow{i} A^0 \xrightarrow{d} A^1 \to \ldots \to A^r \xrightarrow{d} A^{r+1} \to \ldots$$

is exact, where d *is the exterior derivative.*

Now the sheaves A^r are fine[180] for $r \geq 0$, and so by the remarks of §10.2 we have immediately <u>de Rham's theorem</u>[181].

166

<u>Corollary 11.5.2</u> *The* r-th *homology group of the chain complex*

$$O \to A^0(X) \xrightarrow{d} A^1(X) \to \ldots \to A^r(X) \xrightarrow{d} A^{r+1}(X) \to \ldots$$

of R-vector spaces, is $H^r(X;R)$.

A <u>Riemannian metric</u> on X is a positive definite symmetric pairing $m : T \otimes T \to D_X$. Such a pairing induces an inner product in the tangent space at x for each $x \in X$ which varies smoothly with x. A manifold with a chosen Riemannian metric is called a <u>Riemannian manifold</u>. Any compact manifold admits[182] a Riemannian metric.

The D_X-dual $m^* : D_X^* \to (T \otimes T)^*$ of the Riemannian metric is naturally a D_X-module homomorphism $m^* : D_X^* \to A^1 \otimes A^1$ (since T and A^1 are dual) and such a homomorphism is determined uniquely by the element $m^*(1) \in (A^1 \otimes_{D_X} A^1)(X) = A^1(X) \otimes_{D_X(X)} A^1(X)$. Conversely, this element, or equivalently (since A^1 is a sheaf) a compatible family of elements $\{s_i = m^*(1) \in (A^1 \otimes_{D_X} A^1)(U_i) : i \in I\}$ when $\{U_i : i \in I\}$ is an open cover of X, determines m. In particular, the manifold X has a covering by affine open sets. Suppose $U \subset X$ is affine, and $x : U \to R^n$ is a diffeomorphism onto its image. If $x_i : U \to R$ is the i-th component of x, then $x_i \in D_X(U)$ determines an element $dx_i \in \Omega_X(U)$, and the elements $dx_1, \ldots, dx_n$ generate $\Omega_X(U)$ as a free $D_X(U)$-module. The Riemannian metric determines an element of $A^1 \otimes_{D_X} A^1(U)$ which can thus be written uniquely in the form $\Sigma g_{i,j} dx_i \otimes dx_j$ where $g_{i,j} \in D_X(U)$, and the $n \times n$ matrix $\|g_{i,j}\|$ is symmetric and positive definite at each point of U.

<u>Example 11.5.2</u> In the unit disc $U = \{(x,y) \in R^2 : x^2 + y^2 < 1\}$ the standard 'flat' Riemannian metric is $dx \otimes dx + dy \otimes dy$. Another Riemannian metric on U is $dx \otimes dx + dy \otimes dy/(1 - x^2 - y^2)$.

Since the Reimannian metric $m : T \otimes T \to D_X$ is non-singular, it induces an isomorphism $h : T \to \mathrm{Hom}_{D_X}(T, D_X) = A^1$, and hence for each r an isomorphism from A^r to $A^{n-r} \otimes K$, where $K = A^n$ is the canonical sheaf, namely the composite

$$A^r \to A^r \otimes (A^n)^* \otimes A^n \to (A^{n-r})^* \otimes K \xrightarrow{\;\lambda^{n-r}(h^{-1}) \otimes 1\;} A^{n-r} \otimes K .$$

Now the manifold X is orientable if and only if $K = D_X$, in which case $** = (-1)^r : A^r \to A^r$, and there is an evaluation map $f \to \int_X f$ from $A^n(X)$ to R defined as follows. Choose a triangulation of X with n-simplices S_i for $i = 1, \ldots, N$, and orient this triangulation. The element $f \in A^n(X)$ defines by restriction an element $f_i \in A^n(S_i)$ which we can consider as an n-differential on an oriented n-simplex in R^n, and which we can then integrate. Now put $\int_X f = \Sigma \int_{S_i} f_i$.

The map $(a,b) \mapsto \int_X a \wedge b$ is a non-singular real pairing from $A^r(X) \otimes A^{n-r}(X)$ R. This is essentially a restatement of Poincaré duality. Now $(a,b) \to \int_X a \wedge * b$ is a non-singular pairing from $A^r(X) \otimes A^r(X) \to R$.

<u>Lemma 11.5.3</u>   *Put* $\delta = - * d * : A^r \to A^{r-1}$. *The operators*
$d : A^r(X) \to A^{r+1}(X)$ *and* $\delta : A^{r+1}(X) \to A^r(X)$ *are adjoint.*

<u>Proof</u>   For $a \in A^r(X)$, $b \in A^{r+1}(X)$ we have

$$d(a \wedge * b) = da \wedge * b + (-1)^r a \wedge * d * b$$

$$= da \wedge * b + a \wedge * * * d * b$$

$$= da \wedge * b - a \wedge * \delta b$$

and so $\langle da, b \rangle - \langle a, \delta b \rangle = \int_X (da \wedge * b - a \wedge * \delta b) = \int_X d(a \wedge * b)$

$$= 0 \quad \text{by Stoke's theorem}^{183}.$$

The operator $\Delta = d\delta + \delta d$ is the <u>Laplace operator</u> (of the Riemannian manifold). Clearly if $da = \delta a = 0$ then $\Delta a = 0$. Conversely if $\Delta a = 0$ then $0 = \langle \Delta a, a \rangle = \langle d\delta a + \delta da, a \rangle = \langle d\delta a, a \rangle + \langle \delta da, a \rangle = \langle \delta a, \delta a \rangle + \langle da, da \rangle$ and so $da = \delta a = 0$. The kernel $B^r(X)$ of $\Delta : A^r(X) \to A^r(X)$ is called the space of <u>harmonic r-forms</u>. The crucial result is <u>Hodge's theorem</u>[184].

__Theorem 11.5.4__   *The space $A^r(X)$ is the direct sum of the three orthogonal components $\mathrm{Im}(d)$, $\mathrm{Im}(\delta)$ and $B^r(X)$.*

It follows that the composite $B^*(X) \to \mathrm{Ker}\ d \to \mathrm{Ker}\ d/\mathrm{Im}\ d$ is an isomorphism.   The latter group is $H^*(X;R)$ by 11.5.1 and so we have an isomorphism $B^*(X) \to H^*(X;R)$ of graded R-modules.   This isomorphism preserves multiplication, and so the inner product on $B^*(X)$ as a subalgebra of $A^*(X)$ induces an inner product in cohomology algebra.

11.6   COMPLEX HODGE THEORY

Let (X,H_X) be a complex manifold, compact and connected, of dimension n .   Let T_{an} be its tangent sheaf and Ω^*_{an} the H_X-algebra of differential forms.   Let (X,D_X) be the underlying smooth real manifold of (X,H_X) .

Consider the natural map $i : (X,D_{M,C}) \to (X,D_M)$ (see §11.1).   If Ω_X is the real cotangent bundle of (X,D_X) then $i^*\Omega_X$ is its complexification. Now if Ω_{an} is the complex cotangent bundle and Ω is the $D_{X,C}$ module $D_{X,C} \otimes_{H_X} \Omega_{an}$, then since $i_*\Omega = \Omega_X$, there is a natural isomorphism $i^*\Omega_X \cong \Omega \oplus \overline{\Omega}$ of $D_{X,C}$ modules by 11.1.1.   Similarly if T_X is the real tangent sheaf of (X,D_X) and $T = D_{X,C} \otimes_{H_X} T_{an}$, then the $D_{X,C}$ modules i^*T_X and $T \oplus \overline{T}$ are naturally isomorphic.

Unless otherwise stated, we shall work in the category $\underline{D}_{X,C}\underline{-\mathrm{Mod}}$. That is to say, we work with differentiable sections of smooth complex vector bundles.   Since $i^*\Omega_X$ is isomorphic to $\Omega \oplus \overline{\Omega}$, there is a natural isomorophism from $i^*A^r = \lambda^r(C \otimes_R A^r)$ to $\underset{p+q=r}{\oplus} A^{p,q}$ where $A^{p,q} = \lambda^p(\Omega) \otimes \lambda^q(\overline{\Omega})$.   Elements of $A^{p,q}(U)$ for $U \subset X$ open are called (p,q)-__forms__ on U .   The operator $1 \otimes d$ can be written as $\partial \oplus \overline{\partial}$ where $\partial : A^{p,q} \to A^{p+1,q}$ and $\overline{\partial} : A^{p,q} \to A^{p,q+1}$ satisfy $\partial^2 = \overline{\partial}^2 = \partial\overline{\partial} + \overline{\partial}\partial = 0$.

There is a natural inclusion of Ω^p_{an} into $A^{p,0}$, and we have a complex analogue of the <u>Poincaré lemma</u>[185].

<u>Lemma 11.6.1</u>   *The sequence of sheaves of complex vector spaces*

$$0 \to \Omega_{an} \to A^{p,0} \xrightarrow{\bar{\partial}} A^{p,1} \to \ldots \to A^{p,q} \xrightarrow{\bar{\partial}} A^{p,q+1} \to \ldots$$

is exact.

Since these sheaves $A^{p,q}$ are fine[186] for $p,q \geq 0$ we have immediately:

<u>Corollary 11.6.2</u>   *The cohomology group* $H^{p,q}(X) = H^q(X, \Omega^p_{an})$ *is the* q-th *homology group of the chain complex*

$$0 \to A^{p,0}(X) \xrightarrow{\bar{\partial}} A^{p,1}(X) \to \ldots \to A^{p,r}(X) \xrightarrow{\bar{\partial}} A^{p,r+1}(X) \to \ldots .$$

We note that $H^{p,0}(X)$ is the complex vector space $\Omega^p_{an}(X)$.   In general, the sheaves Ω^p_{an} are not fine.

A <u>Hermitian metric</u> on X is a symmetric positive definite Hermitian pairing $m : T \otimes \bar{T} \to D_{X,C}$.   As in the real case such a pairing induces by duality a map $m^* : D^*_{X,C} \to (T \otimes T)^* = A^{1,1}$ which is determined locally.   If $U \subset X$ is an affine open set, and $z : U \to C^n$ is a holomorphic chart with components $z_i : U \to C$,   then $dz_1, \ldots, dz_n$ form a basis for the $H_X(U)$ module $\Omega_{an}(U)$,   and hence the $D_{X,C}$-module $\Omega(U)$.   Let $d\bar{z}_1, \ldots, d\bar{z}_n$ be the same elements of $\bar{\Omega}(U)$.   Locally, the Hermitian metric can be expressed as $\Sigma g_{r,s} dz_r \otimes d\bar{z}_s$ where $g_{r,s} : U \to C$ is a (real-)differentiable function such that the $n \times n$ complex matrix $\|g_{r,s}\|$ is a positive definite Hermitian matrix at each point of U.

<u>Example 11.6.3</u>   If $U = \{z \in C : |z| < 1\}$ is the unit disc, then $dz \otimes d\bar{z}/(1 - |z|^2)$ is a Hermitian metric, called the Poincaré metric.   This metric is[187] invariant under the holomorphic automorphisms of U.

Any compact complex manifold[188] admits a Hermitian metric.   This induces an isomorphism $A^{p,q} \xrightarrow{\sim} A^{n-q,n-p}$ namely the composite

$$A^{p,q} = \lambda^p\Omega \otimes \lambda^q\bar{\Omega} \to \lambda^p\Omega \otimes \lambda^n\Omega^* \otimes \lambda^n\Omega \otimes \lambda^q\Omega^* \to \lambda^{n-p}\Omega^* \otimes \lambda^{n-q}\Omega \to \lambda^{n-q}\Omega \otimes \lambda^{n-p}\bar{\Omega} = A^{n-q,n-p}.$$

As before, we have $** = (-1)^{p+q} : A^{p,q} \to A^{p,q}$, and $A^{p,q}(X)$ has a Hermitian inner product given by $\langle a,b \rangle = \int_X a \wedge *b$. We put $\theta = -*\bar{\partial}*$. As in the real real case, $\bar{\partial}$ and θ are adjoint, and the operator $\square = \bar{\partial}\theta + \theta\bar{\partial}$ is called the Laplace-Beltrami operator. The kernel of $\square$ is the intersection of $\text{Ker}(\bar{\partial})$ and $\text{Ker}(\theta)$ and $B^{p,q}(X) = \text{Ker }\square : A^{p,q}(X) \to A^{p,q}(X)$ is called the space of harmonic (p,q) forms. The following result is Hodge's theorem.

Theorem 11.6.4 *The space $A^{p,q}(X)$ is the direct sum of the three orthogonal components $\text{Im}(\bar{\partial})$, $\text{Im}(\theta)$ and $B^{p,q}(X)$.*

We have an immediate corollary.

Corollary 11.6.5 *The map $B^{p,q}(X) \to \text{Ker}(\bar{\partial}) \to \text{Ker}(\bar{\partial})/\text{Im}(\bar{\partial}) = H^{p,q}(X)$ is an isomorphism of graded C-algebras.*

In particular, $H^{*,*}(X)$ is a complex normed algebra and, since[190] the Laplace-Beltrami operator is an elliptic differential operator, is finite dimensional.

11.7 KAHLER MANIFOLDS

Let (X,H_X) be a complex manifold, compact and connected, with underlying real manifold (X,D_X). If T denotes the $D_{X,C}$-module $C \otimes T_{an}$ and $m : T \otimes_{D_{X,C}} T \to D_{X,C}$ is a Hermitian metric, then the composite

$$T \otimes_{D_X} T \to T \otimes_{D_{X,C}} T \to D_{X,C} \xrightarrow{\text{Re}} D_X$$

(where Re is the map taking the real part of a complex valued function) is a Riemannian metric on (X,D_X). The composite $A^{1,1} \to \otimes A^{p,q} \to A^2$ associates to the element $\omega = m^*(1)$ $A^{1,1}(X)$ an element τ of $A^2(X)$.

If U is a complex affine open set and $z : U \to C^n$ is a chart with coordinates $z_r : U \to C$, and x_r, y_r are the real and imaginary parts of z_r, then $z_r \in H_X(U) \subset D_{X,C}(U)$ is equal to $x_r + iy_r$, and $dz_r = dx_r + idy_r$

while $d\bar{z}_r = dx_r - idy_r$.   If $\omega = \Sigma\, g_{r,s}\, dz_r \otimes d\bar{z}_s$ in U, then

$$\tau = \Sigma\, g_{r,s}(dx_r \wedge dx_s + dy_r \wedge dy_s) - i\Sigma\,(g_{r,s} + g_{s,r})(dx_r \wedge dy_s).$$

Since $g_{r,s} = \bar{g}_{s,r}$ we see that this complex 2-form has no real part, and so $i\tau$ is a real 2-form, and lies in $A^2(X)$.

The complex manifold $P^n(C)$ has a 'natural' Hermitian metric as follows. If $U_j = \{(x) \in P^n(C) : x_j \neq 0\}$ then the function $z_r : U_j \to C$, $(x) \to x_r/x_j$ is holomorphic.   If $\dot{\omega}_j$ is the element $\underset{r,s\neq j}{\Sigma}\, g_{r,s}\, dz_r \otimes d\bar{z}_s \in A^{1,1}(U_j)$ where $g_{r,s} = \dfrac{\partial^2}{\partial z_r \partial z_s}\, \log \underset{r,s\neq j}{\Sigma}\, z_r \bar{z}_s$,   the the elements ω_j for $j = 0,\ldots,n$ are compatible and so determine an element $\omega \in A^{1,1}(X)$ whose restriction to U_j is ω_j, and so which corresponds to a Hermitian metric on $P^n(C)$.   A simple calculation shows that the associated real 2-form $i\tau$ is harmonic, and that the corresponding element of $H^2(X;R) = B^2$ is the class of a hyperplane.

If X is a complex submanifold of $P^n(C)$ then this Hermitian metric on $P^n(C)$ induces[191] a Hermitian metric on X with the property that the associated real 2-form is again harmonic.   Any Hermitian metric on a complex manifold with this property is called a <u>Kähler metric</u>, and a <u>Kähler manifold</u> is a complex manifold with a chosen Kähler metric.

The fundamental property of a Kahler metric[192] is that the Laplace-Beltrami operator $\Box : A^{*,*} \to A^{*,*}$ is precisely $\Delta/2 : C \otimes A^* \to C \otimes A^*$ where $\Delta : A^* \to A^*$ is the Laplace operator with respect to the underlying Riemannian metric.   It follows that the complexification $C \otimes B^*$ of the algebra of real harmonic forms is isomorphic (naturally) to the algebra $B^{*,*}$ of complex harmonic forms, and so by 11.5.4 and 11.6.5, $H^*(X;C)$ is isomorphic to $H^{*,*}(X)$ and so inherits an inner product making it a complex normed algebra.

Remark 11.7.1   Since $\square$ is a real operator on a Kähler manifold, it commutes with conjugation $A^{p,q} \to A^{q,p}$ and so induces an isomorphism $B^{p,q} \to B^{q,p}$.   For such a manifold, it follows that the odd-dimensional Betti numbers are even.   There are complex manifolds which do not have this property[193], and so are not algebraic manifolds.

11.8   THE RIEMANN HYPOTHESIS FOR ALGEBRAIC MANIFOLDS

An analogue of the Riemann hypothesis would (cf. 10.4.2) relate the absolute values of the eigenvalues of the endomorphisms induced on cohomology by an algebraic map to its topological degree.   In general, there may be no relation.

Example 11.8.1   If V is the product $C/Z^2 \times C/Z^2$ of two elliptic curves, then V is an algebraic manifold. For any two integers a,b the map $(x,y) \mapsto (ax,by)$ from V to V is algebraic and has topological degree $D = a^2 b^2$, but the eigenvalues of the endomorphism induced on $H^2(V;R)$ are a^2, ab and b^2.   Thus the cohomological interpretation of the ζ-function must depend on some additional properties of the Frobenius map.

Suppose K is the algebraic closure of F_q and $V \subset P^n(K)$ is a smooth irreducible K/F_q-variety.   If H is a hyperplane in $P^n(K)$ then since $\pi^*(H) = qH$ in $C^1_{alg}(P^n(K))$ by 9.11.3   , we see that $\pi^*(D) = qD$ in $C^1_{alg}(V)$, where D = H.V.   The following analogy of the Riemann hypothesis is due to Serre[194].

Theorem 11.8.2   *If* $X \subset P^n(C)$ *is a smooth connected subvariety and* $\pi:X \to X$ *is an algebraic isomorphism such that* $\pi^*(D) = qD$, *where* $D \in C^1_{alg}(V)$ *is the class* $X.P^{n-1}(C)$, *then the eigenvalues of the endomorphism* $H^r(\pi;C)$ *of* $H^r(X;C)$ *have absolute value* $q^{r/2}$.

__Proof__   The Kähler metric induced on X has fundamental class d in $H^2(X;R)$ equal to the class cl(D) of the divisor D, and the equation $\pi^*(D) = qD$ in $C^1_{alg}(X)$ implies $\pi^*(d) = qd$ in $H^2(X;R)$.   Now the image of the fundamental class in $H^2(X;R)$ belongs to $H^{1,1}(X)$ since it is the class of the metri $\omega \in B^{1,1}$.

The map π is holomorphic and induces an endomorphism of the sheaves $A^{*,*}$ which commutes with $\square$.   Now $\pi^*(\omega) - q\omega \in Im(\bar{\partial})$, and so by Hodge's theorem 11.6.4 we have $\pi^*(\omega) = q\omega$ since ω is harmonic.   If we put $f^{p,q} = q^{-r/2}\pi^{p,q}:A^{p,q} \to A^{p,q}$, where r = p+q, the $f^{*,*}$ is a continuous algebra homomorphism on $A^{*,*}$ preserving the metric and hence the inner product.   In particular $f^{*,*}:B^{*,*} \to B^{*,*}$ is unitary, so its eigenvalues have norm 1 and the result follows.

11.9   THE RIEMANN HYPOTHESIS FOR CURVES

Let K be an algebraically closed field, and let $V \subset P^n(K)$ be a smooth irreducible curve.   A __correspondence__ on V is a divisor on V x V, and the group $C^1_{alg}(V \times V)$ of classes of correspondences is denoted by Cor(V). If A,B are two divisors on V x V such that A x V and V x B intersect properly on V x V x V, then their composite AB is the correspondence $p_{1,3*}(A \times V.V \times B)$, where $p_{i,j}:V \times V \times V \to V \times V$ is the projection $(v_1,v_2,v_3) \mapsto (v_i,v_j)$.   This operation extends by virtue of the Moving lemma to a product in Cor(V),   namely $ab = p_{1,3*}(p_{1,2}{}^*a.p_{2,3}{}^*b)$, making Cor(V) into a ring.   The identity of Cor(V) is the algebraic equivalence class of the diagonal Δ.   The elements of this ring determined by the correspondences v x V and V x v for some point $v \in V$ are denoted by i_1 and i_2.

For any $a \in \mathrm{Cor}(V)$, the element $S(a) = a.(i_1 + i_2 - 1)$ belongs to $C^2_{\mathrm{alg}}(V \times V) = Z$ and the function $a \mapsto S(a)$ is an additive homomorphism from $\mathrm{Cor}(V)$ to Z which vanishes on the ideal I generated by i_1 and i_2, and so induces an additive homomorphism σ from the ring $A = \mathrm{Cor}(V)/I$ to Z. Each divisor D on $V \times V$ has a transpose $D' = \{(v,v') \in V \times V : (v',v) \in D\}$, and the function $D \mapsto D'$ induces an automorphism $a \mapsto a'$ of A such that $\sigma(a) = \sigma(a')$.

The important theorem is the following[195].

Theorem 11.9.1

(11.9.1.1) *The integer $\sigma(1)$ is* 2g, *where* g *is the genus of* V.

(11.9.1.2) *For any* $a \in A$ *we have* $\sigma(aa') \geq 0$.

The inequality (11.9.1.2) is known as Castelnuevo's formula[196], and (11.9.1.1) is a consequence of the Riemann–Roch theorem[197]. The proof of the Riemann hypothesis for curves now follows from 11.9.1 as we now indicate.

Suppose now $K = F_q^a$, and let $\pi \in A$ be the class of the graph of the Frobenius map π from V to V; the element π^n of A is then determined by the graph of π^n. Proof of the following is quite straightforward[198].

Lemma 11.9.2 *We have*

(11.9.2.1) $\sigma(\pi^n \pi'^n) = q^n$

(11.9.2.2) $\sigma(\pi^n) = 1 + q^n - \nu_n$ *where* ν_n *is the number of fixed points of* π^n.

Now for any $x, y \in Z$ let a be the element $x + y\pi^n$ of A so that

$$\sigma(aa') = \sigma(x^2 + xy\pi^n + xy\pi'^n + y^2\pi^n\pi'^n)$$

$$= x^2\sigma(1) + 2xy\sigma(\pi^n) + y^2\sigma(\pi^n\pi'^n)$$

$$= 2gx^2 + 2(1 + q^n - \nu_n)xy + 2q^n y^2$$

using 11.9.1 and 11.9.2. Since $\sigma(aa') \geq 0$ for all values of x,y we have

(11.9.3) $$|1 + q^n - \nu_n| \leq 2gq^{n/2}.$$

Put $P(t) = (1 - t)(1 - qt)Z(t)$, where $Z(t)$ is the meromorphic function $\exp \Sigma \nu_n t^n/n$. Since $t\frac{d}{dt} \log P(t) = \Sigma (1 + q^n - \nu_n)t^n$, we see by the comparison test using (11.9.3) that $\log P(t)$ converges in the disc $U = \{t \in C : |t| \leq q^{\frac{1}{2}}\}$, and so $P(t)$ has no zeros in U, and hence neither does $Z(t)$. From the functional equation (5.5.3) we see that $Z(1/qt)$ has no zeros in U, and so the zeros of $Z(t)$ lie on ∂U and have absolute value $q^{\frac{1}{2}}$.

<u>Remark 11.9.4</u>   It is interesting to note that this proof of the Riemann hypothesis , essentially for global fields on non-zero characteristic, exists outside the category of such fields since it makes use of varieties of dimension greater than 1.

We now indicate with the aid of cohomology and Hodge theory how complex analogue of 11.9.1 may be proved. Suppose now that V is a smooth complex projective curve of genus g.

Consider the class map $cl_{V \times V}: C^1_{alg}(V \times V) = Cor(V) \to H^2(V \times V; R)$. By naturality (10.3.3) we see that for any $a \in A$, the integer $S(a)$ is precisely $cl_{V\times V}(a).(i_1 + i_2 - \Delta) \in H^4(V \times V; R) = R$, where i_1, i_2 and Δ are the fundamental classes of $v \times V$, $V \times V$ and the diagonal. The Kunneth formula[199] shows that $H^2(V \times V; R)$ is isomorphic to $H^1(V;R) \oplus (E \otimes_R E) \oplus H^1(V;R)$, where $E = H^1(V;R)$ and the two components of $H^2(V;R)$ are generated by i_1 an i_2.

If for any element $a \in Cor(V)$ we define M_a to be the component of $cl_{V \times V}(a)$ in $E \otimes_R E$, then $a \mapsto M_a$ factors through A, and if $S':E \otimes_R E \to R$ is the map induced by $e_1 \otimes e_2 \mapsto e_1 e_2(i_1 + i_2 - \Delta)$ (identifying $H^4(V \times V; R)$ with R by Poincaré duality), then $S(a) = S'(M_a)$. If we also use Poincaré duality to identify E with its dual $Hom_R(E,R)$, then we can identify $E \otimes_R E$ with $End_R(E)$ by means of the isomorphism

176

$$E \otimes_R E \to E \otimes_R \mathrm{Hom}_R(E,R) \to \mathrm{Hom}_R(E,E) = \mathrm{End}_R(E)$$

then S' is precisely the trace homomorphism.   It is easy to see by naturality

of the class map that $a \mapsto M_a$ from the ring A to End (E) is a ring

homomorphism.

The Riemann surface V_{an} is a Kähler manifold and so there are

isomorphisms $H^1(V;R) \otimes C \to B^1 \otimes C \to B^{1,0} \oplus B^{0,1}$.   Suppose $e \in E = H^1(V;R)$

is any element.   This can be expressed uniquely as z + w where $z \in B^{1,0}$ and

$w \in B^{0,1}$, or, since conjugation is an isomorphism from $B^{1,0}$ to $B^{0,1}$, as

$z + \bar{z}'$ where $z,z' \in B^{1,0}$.   Since e is real, $e = \bar{e}$, and so z = z' and

$e = z + \bar{z}$ is the real part of a unique complex harmonic form.   In other

words, the composite $E \to C \otimes E \to C \otimes B^1 \to B^{1,0} \oplus B^{0,1} \to B^{1,0}$,

where the last map is projection, is an isomorphism and so E is naturally

a <u>complex</u> vector space,   Equivalently, if $e \in E$ is the real part of the

complex harmonic 1-form z and we put J(e) to be the imaginary part of z

then $J:E \to E$ is C-linear and satisfies $J^2 = -1$, so induces on E a complex

structure.   The map $H^1(\pi;R):E \to E$ induced by a holomorphic map $\pi:V \to V$

is complex linear, and more generally for any $a \in A$, the map $M_a:E \to E$

is complex linear.

Choose a base point $\star \in V$.   If $f:I \to V$ is a path in V with f(0)=$\star$

then any element e of $E = B^{1,0}$ can be integrated along f to give a complex

number $\int_f e$.   The map $e \mapsto \int_f e$ from E to C is C-linear, so that

$\int_f \in E^* = \mathrm{Hom}_C(E,C)$.   Those functionals which correspond to loops at the

base point form a lattice[200] in E^* isomorphic to $H_1(V;Z)$,   the

abelianisation of $\pi_1(V,\star)$.   The quotient complex torus $J'(V) = E^*/H_1(V;Z)$

is precisely the Jacobian J(V).   For if v is any point of V and we

choose a path f in V from $\star$ to v , then the element $\int_f e \in E$

determines an element h(v) $\in J'(V)$ which is independent of the choice of path, and defines a holomorphic function $h:V \to J'(V)$. The function $V^g \to J'(V)$, $(x_1,\ldots,x_g) \mapsto h(x_j)$ can be shown to induce an isomorphism from the g-fold symmetric product $J(V)$ of V to $J'(V)$, by using the theory of Abelian integrals.[201]

<u>Remark 11.9.5</u> In particular if $g = 1$ so that V is an elliptic curve, then the projection map $E^* \to J'(V) = J(V) = V$ is the universal covering space, and $H_1(V;Z) \subset E^*$ is the orbit of 0 under the action of $\pi_1(V,*)$. This indicates the cohomological significance of Hasse's proof (see §8.3).

We now prove the complex analogue of 11.9.1. The identification of E with $B^{1,0}$ induces on E a Hermitian inner product. One can show[202] that for any element $a \in A$, the elements M_a and $M_{a'}$ of $\text{End}_C(E)$ are adjoint, and so $\sigma(aa') = \text{Tr}(M_{aa'}) = \text{Tr}(M_a M_{a'})$ which is thus non-negative. The equation $\sigma(1) = 2g$ is simply $\text{Tr}(\text{id}_E) = \dim_R(E) = 2g$.

<u>Remark 11.9.6</u> There is[203] a purely algebraic analogue of this cohomological proof which essentially uses ℓ-adic cohomology and which applies equally to the case of a projective curve over a finite field of characteristic not equal to ℓ. In this version, the real vector space $H^1(V;R)$ is replaced by the Q_ℓ-vector space $Q_\ell \otimes_{Z_\ell} T_\ell$, where T_ℓ is the <u>Tate module</u> $\varprojlim\{\text{Ker } \ell^n:J(V) \to J(V)\}$ which in the complex case is isomorphic to $H^1(J(V);Z_\ell)$.

12 Schemes

12.1   SCHEMES

Just as in Chapter 5 the concept of global field encompasses both algebraic number fields and fields of rational functions on a curve, so the concept of a scheme is a simultaneous generalisation of a ring and a variety.

For any ring R, the set $\mathrm{spec}(R)$ of prime ideals of R carries a natural algebraically defined topology. If I is an ideal of R, let $V(I) = \{P \in \mathrm{spec}(R) : I \subset P\}$. It is easy to see that $V(\Sigma I_\alpha) = \cap \, V(I_\alpha)$ and that $V(I \cap J) = V(I) \cup V(J)$. The topology with the sets $V(I)$ as closed sets is called the $\underline{\text{Zariski}}$ topology. Clearly the closed points of $\mathrm{spec}(R)$ are the points of $\mathrm{max}(R)$.

If X is a closed subspace of $\mathrm{spec}(R)$, the subset $I(X) = \{r \in R: r \in P \quad \forall P \in X\}$ is an ideal of R such that $VI(X) = X$. As for varieties, we have $\underline{\text{Hilbert's Nullstellensatz}}$:[204]

$\underline{\text{Theorem 12.1.1}}$ *For any ideal* J *of* R *we have* $IV(J) = \sqrt{J}$.

Thus I and V are inverse bijections between the set of closed subsets of $\mathrm{spec}(R)$ and the set of radical ideals of R. It follows that $\mathrm{spec}(R)$ is quasi-compact, and Noetherian if R is Noetherian (cf. §9.1). For any $f \in R$, put $D(f) = \{P \in \mathrm{spec}(R) : f \notin P\} = \mathrm{spec}(R) \setminus V(\langle f \rangle)$. Since $D(f) \cap D(g) = D(fg)$, the set $D = \{D(f) : f \in R\}$ is a basis for the topology.

$\underline{\text{Example 12.1.2}}$ Suppose V is an irreducible affine K/k – variety. The function $v \mapsto p_v$ (see §9.3) from V to $\mathrm{spec}(k[V])$ is continuous.

When $k = K$, this map is a homeomorphism onto $\max(K[V])$ (see 9.3.2). If $W \subset V$ is an irreducible subvariety, then the kernel of the restriction map $K[V] \to K[W]$ is a prime ideal, say p_W, of $K[V]$ and the function $W \mapsto p_W$ is a bijection from the set of irreducible subvarieties of V to $\operatorname{spec} K[V]$.

<u>Example 12.1.3</u> Suppose X is a compact Hausdorff space, and R is the ring of continuous real-valued functions on X. Each point $x \in X$ determines a maximal ideal m_x of R, namely the kernel of the evaluation map $f \mapsto f(x)$ at x. The function $x \mapsto m_x$ from X to $\max(R)$ is easily shown[205] to be bijective. Now if $f \in R$, then $h^{-1}(D(f)) = \{x \in X : f(x) \neq 0\}$ is an open subset of X, and so h is continuous. Since X and $\max(R)$ are compact Hausdorff spaces, h is a homeomorphism.

If $f : R \to S$ is a ring homomorphism and P is a prime ideal of S then $f^{-1}P$ is a prime ideal of R. The function $f^{-1} : \operatorname{spec}(S) \to \operatorname{spec}(R)$ is clearly continuous, but in general does not preserve the maximal ideals (consider the inclusion $Z \to Q$ for example). This is one reason for working with prime rather than maximal ideals. If I is an ideal of R, the natural map $R \to R/I$ induces a homeomorphism $\operatorname{spec}(R/I)$ to $V(I) \subset \operatorname{spec}(R)$. In particular, if $N = \sqrt{0}$ is the nilradical of R, then $\operatorname{spec}(R/N) \to \operatorname{spec}(R)$ is a homeomorphism.

For $f \in R$, let R_f denote the localisation of R with respect to the multiplicatively closed set of powers of f. We wish to define a sheaf O_R on $\operatorname{spec}(R)$ such that $O_R(D(f)) = R_f$. Since D is a basis for the topology on $\operatorname{spec}(R)$ such a functor is determined uniquely by its restriction to the subcategory D of $\operatorname{Ouv}(\operatorname{spec}(R))$. Suppose $D(f) \supset D(g)$. By 12.1.1, this implies that $\sqrt{\langle f \rangle} \supset \sqrt{\langle g \rangle}$, and so $g^n = af$ for some $n \in N$ and $a \in R$. The ring homomorphism $R_f \to R_g$; $c/f^m \mapsto ca^m/g^{nm}$ is independent of the choice

180

of n and a .   Now if $U \subset \mathrm{spec}(R)$ is open, let $O_R(U)$ be the equaliser
of

$$\prod_{D(f) \subset U} R_f \longrightarrow\longrightarrow \prod_{D(fg) \subset U} R_{fg}$$

so that O_R is a presheaf of rings on $\mathrm{spec}(R)$.   It is[206] in fact a sheaf
whose stalk $O_{R,P}$ at the prime ideal $P \in \mathrm{spec}(R)$ is

$$\varinjlim O_R(U) = \varinjlim \{R_f : f \notin P\} = R_P,$$ the localisation of R at P.

This construction is natural, so we have a functor

$\mathrm{spec} : \underline{\mathrm{Rings}} \to \underline{Z\text{-}\mathrm{Top}}_O$.   Since $O_R(\mathrm{spec}(R)) = R$ this functor is faithful.
It is also full, and so it satisfies (9.6.2).   The corresponding subcategory
(see §9.6) of affine objects in $\underline{Z\text{-}\mathrm{Top}}_O$ is called the category of <u>affine</u>
<u>schemes</u>, and is isomorphic to the <u>opposite</u> category of the category of rings.

<u>Remark 12.1.4</u>   We shall often identify a ring R with the corresponding
affine scheme $\mathrm{spec}(R)$.   However, we must remember that a homomorphism
$f : R \to S$ of rings is a morphism $f : S \to R$ of affine schemes.

The category <u>Rings</u> has a sum (tensor product over Z).

<u>Definition 12.1.5</u>   The category <u>la(spec)</u> is the category of <u>Presch</u> of
<u>preschemes</u>, and the category <u>Sch</u> of separated preschemes is the category
of <u>schemes</u>.

<u>Remark 12.1.6</u>   Some authors use the words scheme/separated scheme instead
of prescheme/scheme.

Any affine scheme is[207] a scheme.   If R is a ring, and $f \in R$ then
then the ring homomorphism $R \to R_f$ induces an isomorphism[208]
$\mathrm{spec}(R_f) \overset{\sim}{=} (D(f), O_R|_{D(f)})$ where O_R is the structure sheaf of $\mathrm{spec}(R)$.
Hence any open subset of an affine scheme is itself a scheme so spec

satisfies (9.6.8).

<u>Example 12.1.7</u>   For any ring R,   the diagram $R[X] \to R[X,X^{-1}] \leftarrow R[X^{-1}]$ of rings induces a diagram

$$\text{spec}\,R[X] \;\leftarrow\; \text{spec}\,R[X,X^{-1}] \;\to\; \text{spec}\,R[X^{-1}]$$

of affine schemes, whose colimit is a scheme $P^1(R)$ called the <u>projective line</u> over R.   In a similar way[209] one can define <u>n-dimensional projective space</u> $P^n(R)$ over any ring.

Since a scheme is locally a ring, various definitions and properties of rings can be generalised to preschemes, either through the values of the structure sheaf on the open sets, or through its stalks.   We give some of those which we shall need.

<u>Definition 12.1.8</u>   A scheme X is

(12.1.8.1) <u>reduced</u> if $O_X(U)$ has no zero-divisors for $U \subset X$ open;

(12.1.8.2) <u>integral</u> if it is irreducible and reduced;

(12.1.8.3) <u>Noetherian</u> (<u>locally-Noetherian</u>) if it is the union of a finite (arbitrary) number of Noetherian affine schemes;

(12.1.8.4) <u>normal</u> if each stalk $O_{X,x}$ is integrally closed.

<u>Definition 12.1.9</u>   A morphism $(f,\alpha) : (X,O_X) \to (Y,O_Y)$ of schemes is <u>of finite type</u> if for each open affine set $U \subset Y$ the set $f^{-1}(U)$ is a finite union of affine open sets $V_1,\ldots,V_r$ such that each $O_X(V_i)$ is a finitely generated $O_Y(U)$ algebra via the map

$$O_Y(U) \xrightarrow{\;\;\alpha\;\;} f_*O_X(U) \;=\; O_X(f^{-1}(U)) \xrightarrow{\hspace{2cm}} O_X(V_i).$$

A prescheme whose base space is a single point is of the form spec(R) where R is a local ring, and so is necessarily a scheme.   Morphisms from such a scheme to a prescheme X correspond to pairs (x,f) where x is a point of X and $f : O_{X,x} \to R$ is a homomorphism of local rings.   Of course

182

when R is a field such a ring homomorphism is determined by the map it induces of the residue field.

<u>Definition 12.2.9</u> If X,Y are preschemes, any Y <u>valued point</u> of X is a morphism $Y \to X$ of preschemes. A (K-valued) <u>geometric point</u> is a K-valued point, where K is an algebracially closed field.

<u>Example 12.1.10</u> If A is the ring of integers in a number field F, then the points of spec(A) are the maximal ideals of A together with the prime ideal O which is not maximal. The residue field k(x) at the point x is A/x if x is maximal, but k(O) = F. The C-valued points of spec(A) correspond to embeddings of F into C, while if k is a field of characteristic $p \neq O$ then the k-valued points of spec(A) correspond to pairs (x,i) where x is a maximal ideal such that k(x) has characteristic p, and $i : k(x) \to k$ is an embedding.

12.2   VARIETIES AS SCHEMES

Suppose $V = (V,O_V)$ is an affine K-variety.   Let $R = K[V] = O_V(V)$ be its ring of regular functions and let sch(V) be the affine scheme spec(R). Consider the natural map $i : v \mapsto m_v$ between the base spaces of V and sch(V).   For any $f \in R$, $O_V(i^{-1}D(f)) = R_f$ so there is a natural morphism $V \to sch(V)$ of local-ringed spaces.   If for any ringed space $X = (X,O_X)$ we write $|X|$ for the ringed space $(Y,O_{X|Y})$ where Y is the set of closed points of X, then the map $V \to sch(V)$ induces an isomorphism $V \to |sch(V)|$ in <u>Z-Top</u>$_O$.   The scheme sch(V) does not by itself determine the variety V, since $V \in Ob(\underline{K\text{-Top}}_O)$.   Now since a K-algebra is a ring R together with a homomorphism $K \to R$ it follows that the category <u>K-Top</u>$_O$ is naturally isomorphic to the category <u>Z-Top</u>$_O$/<u>K</u> of   local-ringed spaces over

K.   That is to say, the variety V is equivalent to the object $V \to K$ in

$\underline{\text{Z-Top}}_0/\underline{K}$.   Of course, the object K in $\underline{\text{Z-Top}}_0$ is at the same time an

affine K-variety (with base space a single point) and an affine scheme

(namely spec(K)).   The morphism $K \to K[V]$ which defines the K-algebra

structure on $K[V]$ induces a morphism $\text{sch}(V) \to K$ of preschemes over K,

and the corresponding object $\left|\text{sch}(V)\right| \to \left|K\right| = K$ of $\underline{\text{Z-Top}}_0/\underline{K} = \underline{\text{K-Top}}_0$ is

naturally isomorphic to V (in $\underline{\text{K-Top}}_0$).

Now the functor $V \mapsto (\text{sch}(V) \to K)$ from the category of affine K-

varieties to $\underline{\text{Presch}/K}$ induces (by gaga, see §9.10) a functor sch from the

category of prevarieties over K to the category $\underline{\text{Presch}/K}$, along with a

natural transformation $V \to \text{sch}(V)$ in $\underline{\text{K-Top}}_0$ which induces an isomorphism

$V \to \left|\text{sch}(V)\right|$.   Moreover, $\text{sch}(V) \to K$ is a reduced scheme of finite type.

Hence we have the following.[210]

<u>Theorem 12.2.1</u> *The functor* sch *from the category of prevarieties over* K

to the category of reduced preschemes of finite type over K *is an*

equivalence.

<u>Notation 12.2.2</u>   If $X \to K$ is a reduced scheme of finite type over K, we

write X(K) for the corresponding K-variety.

For any prescheme S, the category $\underline{\text{Presch}/S}$ has products denoted by

$(X,Y) \mapsto X \times_S Y$. An object $X \to S$ is <u>separated</u> if the image of X under the

diagonal map $X \to X \times_S X$ is closed.   It is clear that under the equivalence

of 12.2.1 the varieties correspond to separated preschemes over K.   Because

of these results, a K-variety could now be redefined[211] as a reduced

separated scheme of finite type over K.

If X,Y are two preschemes over S then a <u>Y-valued point</u> of X is a

morphism $Y \to X$ in $\underline{\text{Presch}/S}$.   In particular a <u>pointed prescheme over</u> S is a pair comprising a prescheme X over S and an S-valued point of X

(where S also denotes the prescheme $1 : S \to S$ over S).

<u>Example 12.2.3</u>   It is easy to see (by considering the above affine case and
the methods of gaga) that if V is a K-variety, then the points of sch(V)
are precisely the irreducible subvarieties of F, and that if $x \in$ sch(V)
is a point corresponding to the irreducible subvariety W of V then the
residue field of sch(V) at x is the field K(W) of rational functions
on W.   Since a K-valued point of sch(V) $\to$ K is a pair (x,i) with
$x \in$ sch(V) and $i : k(x) \to K$ an embedding over K, we see that the K-valued
points correspond to the closed points of sch(V), i.e. to the points of V.

12.3   BASE CHANGE

If $f : S' \to S$ is a map of preschemes and $X \to S$ is a prescheme over S,
then the fibre product $X \times_S S'$ exists and the projection $X \times_S S' \to S'$ is
a prescheme over S', so that f induces a functor from <u>Presch/S</u> to
<u>Presch/S'</u> called <u>base change along</u> f.   The prescheme $X \times_S S'$ is called
the <u>fibre</u> of $X \to S$ along f.   A <u>geometric fibre</u> of $X \to S$ is a fibre
along a geometric point.

<u>Example 12.3.1</u>   The fibre of the map $P^n(\mathbb{Z}) \to \mathbb{Z}$ along the morphism $R \to \mathbb{Z}$
of schemes (corresponding to the map $\mathbb{Z} \to R$ of rings) is precisely $P^n(R)$.
The geometric fibres of $P^n(\mathbb{Z})$ are the n dimensional projective spaces
over algebraically closed fields.

   Base change preserves certain properties of preschemes such as being
Noetherian or of finite type.[212]

<u>Example 12.3.2</u>   Let V be a K-variety and L/K an extension of K with

L   algebraically closed.   The fibre product $sch(V) \times_K L \to L$ is reduced
and of finite type.   By the universal property of fibre product, the L-
valued points of $sch(V) \to K$ correspond bijectively with the L-valued points
of $sch(V) \times_K L \to L$,   hence the points of the L-variety $(sch(V) \times_K L)(L)$.

For example, suppose $V \subset A^n(K)$ is an affine variety and $X \to K$ is
$sch(V) \to K$.   The polynomials over   K   defining   V   considered as polynomials over   L   define an affine variety $V_L \subset A^n(L)$.   The associated
scheme $sch(V_L) \to L$ is $X \times_K L \to L$.

Base change also enables us to discuss K/k-varieties in the language of
schemes.   If   V   is an affine   K/k   variety, then $\operatorname{spec} k[Y]$ is[213]   a
separated reduced scheme of finite type over   k   such that

$\operatorname{spec} k[V] \times_k K = \operatorname{spec} K[V]$.

<u>Definition 12.3.3</u>   A prescheme $V \to K$ is a   K/k   <u>prevariety</u> or is <u>defined</u>
<u>over</u>   k   if there is a reduced scheme of finite type $V' \to k$ such that   V
is the fibre of $V' \to k$ along $K \to k$.

If α is an automorphism of   K/k,   then α induces an automorphism of
the K/k-variety $V' \times_k K$,   namely $id \times_k \alpha$.

Further, base change allows us to formulate more precisely the concepts
of reduction and lifting of varieties.   For example, suppose   K   is an
extension of Q and   S   is a discrete spot on   K   with valuation ring   R
and residue field   k.   Suppose that $X \to R$ is a reduced scheme of finite
type over   R.   Each geometric fibre of   R   is a prevariety over the corresponding field.   For example, if $C \to R$ is a complex valued point of   R,
then $X \times_R C$ is a C/K-prevariety, while if $k' \to k$ is a geometric point of
k   then $X \times_R k'$ is a k'/k   prevariety.   If in particular   k   and   K   are
separably algebraically closed then $X \times_R K$ is called a <u>lifting of</u> $X \times_R k$

to <u>characteristic</u> 0, and $X \times_R k$ is a <u>reduction</u> of $X \times_R K$ <u>mod p</u>, where p is the characteristic of k.

12.4 SMOOTH AND ÉTALE MAPS

The definition of a smooth map between schemes is based on the implicit function theorem.

<u>Definition 12.4.1</u> Let R be a ring and let S be the quotient ring $R[X_1,\ldots,X_{n+m}]/\langle f_1,\ldots,f_n \rangle$ where $f_1,\ldots,f_n$ are polynomials over R. The natural map $R \to S$ is <u>smooth at the prime ideal</u> $P \in \operatorname{spec}(S)$ <u>(of relative degree</u> $m)$ if the Jacobian matrix $J = \| \partial f_i/\partial X_j \|$ has rank n over the quotient ring S/P. The map $R \to S$ is <u>smooth</u> if it is smooth at each prime ideal of S.

Since the categories of rings and of affine schemes are equivalent we have a definition of smoothness for maps between affine schemes. A map between schemes is smooth if it is locally smooth:

<u>Definition 12.4.2</u> A map $f : X \to Y$ of schemes is <u>smooth of relative degree</u> m if for each point $x \in X$ there is an affine open set S containing x and an affine open set R containing $f(x)$ such that $f : S \to R$ is a smooth map of affine schemes of relative degree m. A map is <u>étale</u> if it is smooth of relative degree 0.

<u>Example 12.4.3</u> An extension $K \to L$ of fields is smooth of relative degree m if and only if L/K is finitely generated, separable, of transcendence degree m and is étale if and only if L/K is a finite separable extension.

<u>Example 12.4.4</u> The ring homomorphism $Z \to Z[X,Y]/\langle Y^2 + X^3 + 1 \rangle = S$ is not smooth. The Jacobian $(3X^2, 2Y)$ vanishes under the map from S to F_3

induced by the map $Z[X,Y] \xrightarrow{e} Z \longrightarrow F_3$, where $e(Y) = 0$, $e(X) = -1$.
However, if A is the localisation of Z with respect to the multiplicatively closed set generated by 2 and 3, then $A \to A[X,Y]/\underline{\le}Y^2 + X^3 + 1 >$ is smooth.

<u>Example 12.4.5</u>   If V is a K-variety then the map $sch(V) \to K$ is smooth of relative degree m if and only if V is a smooth variety of dimension m.

The composite of two smooth (étale) maps is[214] smooth (étale).   The category of schemes and étale maps is denoted by <u>Et</u>.   The properties of being smooth or étale are clearly preserved by base change, so if $f : X \to S$ is a smooth map of relative degree m and $K \to S$ is a geometric point, then the fibre $X \times_S K \to K$ is a smooth variety of dimension m.

<u>Example 12.4.6</u>   The scheme $P^n(Z) \to Z$ is a smooth scheme of finite type so its geometric fibres are smooth varieties.   By 12.3.1 we know that its geometric fibres are $P^n(K)$ for K algebraically closed.

<u>Example 12.4.7</u>   Let A denote the Dedekind domain of 12.4.4, and let f be the homogeneous polynomial $f(X_0,X_1,X_2) = X_0 X_2^2 + X_1^3 + X_0^3$ over A.   If $\{i,j,k\} = \{0,1,2\}$ then for each i let A_i be the ring $A[X_j/X_i, X_k/X_i] / \,<f_i>$ where $f_i = f(X_0/X_i, X_1/X_i, X_2/X_i)$.   The map $A \to A_i$ is smooth.   For each pair i,j let $A_{i,j}$ be the ring $A[X_k/X_i, X_j/X_i, Z_k] / \,<f_i, \ Z_k(X_j/X_i) >$.   The ring homomorphisms $A_i \to A_{i,j}$ are smooth.   The colimit of the diagram $spec\,A_{i,j} \rightrightarrows spec\,A_i$ is a scheme X over A which is smooth and of finite type.   Now A has geometric points of all characteristics except 2 and 3.   If $K \to A$ is a geometric point, then the K-variety $X \times_A K$ is the elliptic curve $\{(x_0;x_1;x_2) \in P^2(K) : f(x_0,x_1,x_2) = 0\}$ which is covered by the affine

188

varieties $((\mathrm{spec}\ A_i)\ x_A\ K)(K) = (\mathrm{spec}\ K\ \otimes_A\ A_i)(K)$ for $i = 1,2,3$.

In particular, the smooth complex variety $X\ x_A\ C$ is a lifting to characteristic 0 of the smooth curve $X\ x_A\ F_p^a$ for any prime $p \neq 2,3$.

Smooth maps can be characterised by their geometric fibres.

<u>Definition 12.4.8</u> A map $(f,\alpha):(X,O_X) \to (Y,O_Y)$ between two ringed spaces is <u>flat</u> if for each $x \in X$ the stalk $O_{X,x}$ is a flat $O_{Y,f(x)}$-module via the map $\alpha_x:O_{Y,f(x)} \to O_{X,x}$.

<u>Theorem 12.4.9</u> *A map* $f:X \to Y$ *of schemes is smooth (étale) iff its geometric fibres are disjoint unions of smooth varieties (points).*

<u>Corollary 12.4.10</u> *A scheme* $f:X \to k$ *of finite type over the field* k *is étale iff* k *is a finite union of fields* k_i *with each* k_i/k *a finite separable extension. The category* <u>Et/k</u> *is isomorphic to the category of separable algebraic semi-simple* k*-algebras.*

If K is an algebraically closed field, and $f:X \to Y$ is a morphism of schemes over K of finite type, one can show[215] that f is étale iff for each closed point $x \in X$, the map $O_{Y,f(x)} \to O_{X,x}$ becomes an isomorphism on completion (compare with 10.1.10).

<u>Proposition 12.4.11</u> *If* X *and* Y *are smooth varieties over the complex numbers, then a map* $f:\mathrm{sch}(X) \to \mathrm{sch}(Y)$ *is étale iff the induced map* $f_{an}:X_{an} \to Y_{an}$ *is a local homeomorphism (in the S-topology).*

<u>Proof</u> This is immediate from 10.1.1.

12.5   GROTHENDIECK TOPOLOGIES

The basic observation behind étale cohomology for schemes is that when X is a topological space, the (Čech) cohomology groups are determined by the category <u>Ouv(X)</u> along with the information concerning which families of objects cover X.   The construction of the cohomology or Čech nerve is then purely formal.   This was first formalised by Grothendieck[216] as follows.

<u>Definition 12.5.1</u>   Let $\underline{C}$ be a category with finite limits.   A <u>Grothendieck topology</u> on $\underline{C}$ comprises for each object X in $\underline{C}$ a set J(X) whose elements are families of objects in $\underline{C/X}$ satisfying the following conditions:

(12.5.1.1)   the identity $1:X \to X$ belongs to $J(X)$;

(12.5.1.2)   if $\{f_i:X_i \to X\} \in J(X)$ and $g:Y \to X$ is a map in $\underline{C}$ then the family $\{f_i':X_i \times_Y Y \to Y\}$ belongs to $J(Y)$, where f_i' is the pull-back of f_i along g;

(12.5.1.3)   if $\{f_i:X_i \to X\} \in J(X)$ and for each i we have an element $\{f_{i,j}:X_{i,j} \to X_i\}$ of $J(X_i)$, then the composite family $\{f_i f_{i,j}:X_{i,j} \to X\}$ belongs to $J(X)$.

The elements of $J(X)$ are called <u>coverings</u> of X.   A <u>site</u> is a category with a (Grothendieck) topology.

<u>Example 12.5.2</u>   Let X be a topological space.   There is a natural topology on <u>Ouv(X)</u> where for U an open set in X, $J(U)$ is the set of coverings of U in the usual sense.   It is easy to see that this is a topology since the fibre product $U_1 \times_U U_2$ is the intersection $U_1 \cap U_2$.   We shall write Ouv(X) for the corresponding site.

If $\underline{C}$ is a site and X is an object of $\underline{C}$, then $\underline{C/X}$ is naturally[217] a site with an 'induced topology' in the obvious way.

<u>Example 12.5.3</u>   Let <u>Top</u> be the category whose objects are topological spaces and whose morphisms are local homeomorphisms.   For $X \in$ Ob(<u>Top</u>), let $J(X)$ be the set of families $\{f_i:X_i \to X\}$ having the property that $X = \bigcup \mathrm{Im}(f_i)$.   This defines[218] a topology on <u>Top</u>.   For any object X in <u>Top</u>, the category <u>Top/X</u> is the category $\underline{X}_{1h}$ (see 9.5), which thus carries the induced topology.   We shall write Top and X_{1h} for the corresponding

sites.

There is an obvious definition of continuous morphism between sites.
For example the 'inclusion' functor $\underline{Ouv(X)} \to \underline{X}_{1h}$ is continuous. For
any category $\underline{C}$, a contravariant functor from $\underline{C}$ to $\underline{Sets}$ is called a presheaf (of sets) on $\underline{C}$, and the category of such presheaves is denoted by $\underline{C}^{\wedge}$.

$\underline{\text{Definition 12.5.4}}$ If $\underline{C}$ is a site, then a presheaf $F:\underline{C} \to \underline{Sets}$ is a $\underline{sheaf}$
if for each object X in $\underline{C}$ and each covering $\{f_i:X_i \to X\}$ in $J(X)$, the
sequence

$$* \longrightarrow F(X) \longrightarrow \prod_i F(X_i) \rightrightarrows \prod_{i,j} F(X_i \times_X X_j)$$

is exact.

The category of sheaves on $\underline{C}$ is denoted by $\underline{C}^{\sim}$, and carries a natural
topology[219] so is itself a site. Any site of the form $\underline{C}^{\sim}$ is called a
(Grothendieck) $\underline{topos}$.

For example, we know (by 9.5.4) that the category $\underline{Ouv(X)}^{\sim}$ is precisely
$\underline{X}_{1h}$. In fact the natural topology on $Ouv(X)^{\sim}$ is such that $Ouv(X)^{\sim}$ and X_{1h}
are isomorphic as sites.

As in the case of $Ouv(X)$, we can more generally consider sheaves of
abelian groups and rings on any site. In particular, let $\underline{Ab(C)}$ be the
category of sheaves of abelian groups on the site $\underline{C}$, which is itself a site
by virtue of the topology induced from the embedding $\underline{Ab(C)} \to \underline{C}^{\sim}$. The
category $\underline{Ab(C)}$ is an abelian category with sufficient injectives[220], inside
which one can apply the techniques of homological algebra. For example,
if 1 is the terminal object of $\underline{C}$, the functor $F \mapsto F(1)$ from $\underline{Ab(C)}$ to
$\underline{Ab}$ is left-exact[221]; its derived functors are denoted by $H*(C;F)$ and
called the $\underline{\text{cohomology groups of}}$ C $\underline{\text{with values in}}$ F.

Now the category $\underline{C}^{\sim}$ is a subcategory of $\underline{C}^{\wedge}$. There is [222] a
reflection functor $\underline{C}^{\wedge} \to \underline{C}^{\sim}$, so if A is an abelian group the constant pre-

sheaf at A determines a sheaf $\underline{A}$ by reflection; the groups $H*(\underline{C};\underline{A})$ are simply

written $H*(\underline{C};A)$, the group A being called the <u>coefficient</u> group.   For

example, if X is a topological space then $H*(Ouv(X);A)$ are the Čech

cohomology groups of X with coefficients in A.

There is a second approach to cohomomolgy through simplicial sets and

nerves due to Verdier.[223] Let 0 denote the initial object of the site $\underline{C}$.

The object X of $\underline{C}$ is <u>connected</u> if $X \neq 0$ and $X = X_1 \amalg X_2$ implies that

precisely one of X_1, X_2 is 0.   The site $\underline{C}$ is <u>locally-connected</u> if every

object can be expressed as a sum of connected objects, and is <u>connected</u>

if it is locally-connected and the terminal object 1 is connected.

The underlying category of a site is distributive[224], that is to say

for any set of maps $\{X_i \to X\}$ and $Y \to X$, the natural map

$\amalg_i (X_i \times_X Y) \to (\amalg_i X_i) \times_X Y$ is an isomorphism.   This implies that the

expression of an object in a connected category as a sum of its 'components'

is essentially unique.   For if $Z = \amalg_i X_i = \amalg_j Y_j$ are two decompositions

into connected objects, then $X_i \times_Z Z = X_i = \amalg (X_i \times_Z Y_j)$ and so there is a

unique j   such that $X_i = X_i \times_Z Y_j$, and so $X_i = Y_j$.   Again the

distributivity implies that maps in $\underline{C}$ preserve components, so there is

a functor $\Pi:\underline{C} \to$ Sets which associates to each object its set of components.

We can now construct the Čech nerve of a site by analogy (see §10.2) with

the case $\underline{C} = Ouv(X).$, as follows.   Since $1 \in Ob(\underline{C})$ is terminal, any

covering $\{f_i:X_i \to 1\}$ is determined by the set of objects $\{X_i:i \in I\}$.

For such a covering and $\underline{i} = (i_0,\ldots,i_n) \in I^{n+1}$, put $X_{\underline{i}} = X_{i_0} \times \ldots \times X_{i_n}$,

and let $U^n = \{X_{\underline{i}}:\underline{i} \in I^{n+1}$ and $X_{\underline{i}} \neq 0\}$.   The family U^n is also a

covering of 1, and there are obvious face and degeneracy maps making U* into

simplicial covering of 1, and hence $\Pi U*$ is a simplicial set.

If $\{V_j : j \in J\}$ is a second covering of 1, then a morphism from V to U is a pair (f,h) where $f : J \rightarrow I$ is a function and for each $j \in J$, $h_j : V_j \rightarrow U_{f(j)}$ is a map in $\underline{C}$. Such a morphism induces a simplial map from V* to U* in $\underline{C}$ and hence a map $\Pi V* \rightarrow \Pi U*$ of simplicial sets. Two morphisms from V to U induce homotopic maps from $\Pi V*$ to $\Pi U*$, and we say that V refines U if there is a morphism from V to U. By virtue of (12.5.1.3) any two coverings have a common refinement, and so we obtain an object $\{ \Pi U* : U \in J(1) \}$ of $\underline{\text{Pro-HSS}}$ indexed by the coverings of 1, directed by refinement. This object of $\underline{\text{Pro-HSS}}$ is called the Čech homotopy type of the site $\underline{C}$.

Any object $X = \{X_i : i \in I\}$ of $\underline{\text{Pro-HSS}}$ has associated cohomology groups $H*(X;A) = \lim_{\rightarrow} H*(X_i;A)$ for any coefficient group A. When X is the Čech homotopy type of a site $\underline{C}$, we write $\check{H}*(\underline{C};A)$ for these groups. In general these will not be the same as $H*(\underline{C};A)$, the reason being that the simplicial coverings obtained by the Čech process are too coarse. However, by introducing the concept of a $\underline{\text{hypercovering}}$ (which is essentially a simplicial object U* in $\underline{C}$ such that U^0 is a covering of the terminal object and for each n, U^n is a refinement of the covering generated by $U^0, \ldots, U^{n-1}$) one obtains a set $HR(\underline{C})$ of hypercoverings of $\underline{C}$, directed by refinement. The object $\Pi \underline{C} = \{U* : U* \in HR(\underline{C})\}$ of $\underline{\text{Pro-HSS}}$ is called the $\underline{\text{Verdier homotopy type}}$ of the site $\underline{C}$. The justification for introducing hypercoverings is the following[225].

<u>Theorem 12.5.5</u> *If $\underline{C}$ is a connected site, there is a natural isomorphism* $\check{H}*(\Pi \underline{C};A) \rightarrow H*(\underline{C};A)$ *for any coefficient group* A.

Of course the homotopy type of a site carries more information than does its cohomomology.

12.6   ETALE COHOMOLOGY

We now come to the example for which this machinery has been introduced. Consider the category $\underline{Et}$ of schemes and étale maps.   We say that a family $\{f_i : X_i \to X\}$ of maps in $\underline{Et}$ is a covering of X if the base space of X is covered by the images of the maps f_i.   If $J(X)$ is the set of coverings of X, then J defines a topology on $\underline{Et}$,   and hence on $\underline{Et/X}$ for any scheme X.   We write X_{et} for the site $\underline{Et/X}$.   If X is locally-Noetherian,   the site X_{et} is connected[226],   and so the cohomology grcups of the pro-object X_{et} (which is called the étale homotopy type of X, and is also denoted by X_{et}) coincide , by 12.5.5, with the cohomology groups of the site X_{et}.   When V is a K-variety,   we shall write V_{et} for $sch(V)_{et}$, which is connected since $sch(V)$ is locally-Noetherian.

It is important to note that the étale homotopy type of the variety V depends on the scheme $sch(V)$ rather than the variety V itself.   For example,   any automorphism of K induces an automorphism of the étale homotopy type.

So now we have a definition of cohomology for varieties over any field. In order to repeat the methods of Chapter 10 for varieties defined over a finite field,   we need to choose our coefficient ring in order that the cohomology of an irreducible variety is finitely generated,   and that for smooth varieties we at least have Poincaré duality and the Kunneth formula. This suggests that the coefficient ring should be a field of zero characteristic.   Suppose the field F is such a candidate.   In order that the ζ-function of a smooth variety can be interpreted via the effect of the Frobenius map on cohomology,   it is necessary that if $H^*(-,F)$ denotes our theory,   and V is a smooth curve of genus g,   the F-vector space $H^1(V;F)$ has dimension 2g.

As we now indicate, this means that F cannot be R , nor hence
any subfield or extension of R.   For suppose that H*(-;R) satisfies these
conditions and that V is an elliptic curve over the field k = F_p^a .   The
map f $\mapsto$ H^1(f) = f* from the ring End(V) to the ring End(H^1(V;R)),
which is isomorphic to the ring $M_2(R)$ of 2 x 2 real matrices, preserves
multiplication.   If a:V x V $\to$ V is the group operation on V and
Δ:V $\to$ V x V is the diagonal map, then the sum f + g in End(V) of the
endomorphisms f,g is a(f x g)Δ , and so

$$(f + g)* \ = \ \Delta*(f \times g)*a*$$

$$= \ \Delta*(f* \oplus g*)a* \quad \text{(by the Kunneth formula)}$$

$$= f* + g*.$$

Thus f $\mapsto$ f* induces a ring homomorphism from $R \otimes$ End(V) to $M_2(R)$.   But
there are examples[227] of elliptic curves V for which $R \otimes$ End(V) is the
ring H of quaternions (compare with §8.3) and there is no ring
homomorphism from H to $M_2(R)$, since both these real vector spaces have
dimension 4 and H is a (non-commutative) field and so any non-trivial
ring homomorphism defined on H must be injective.

However, the ring End(V) does have[228] 2-dimensional representations
over the ℓ-adic number field Q_ℓ where ℓ is a prime not equal to p.

<u>Definition 12.6.1</u>   The <u>ℓ-adic cohomology</u> of a variety V is

$$H^*_{et}(V;Q_\ell) \ = \ Q_\ell \otimes_{Z_\ell} \lim H^*(V_{et};Z/\ell^n Z)$$

It is not necessarily true that $H^*_{et}(V;Q_\ell) = H^*(V_{et};Q_\ell)$.

12.7   COMPARISON THEOREMS

There are several results linking the cohomology groups of those sites which
are of interest to us. The first of these compares the étale cohomology of
a smooth complex variety with the classical cohomology of the associated

complex manifold.

If X is a smooth scheme of finite type over C, let X_{an} be the associated complex manifold (with the S-topology). If $f:Y \to X$ is an étale map, then Y is also a smooth scheme of finite type over C (by me ns of the composite $Y \to X \to C$) and f induces a holomorphic map $f_{an}:Y_{an} \to X_{an}$ which is a finite covering space of its image. The following theorem is often called the <u>Riemann-Enriques theorem</u>[229] and is the generalisation of the results of §7.8.

<u>Theorem 12.7.1</u> *The functor* $(f:Y \to X) \mapsto (f_{an}:Y_{an} \to X_{an})$ *is an isomorphism from the category of étale coverings of* X *to the category of finite covering spaces of* X_{an}.

Thus any étale covering $\{f_i:X_i \to X\}$ of the scheme X, determines a family $\{f_{i,an}:X_{i,an} \to X_{an}\}$ of local homeomorphisms into X_{an}, i.e. a covering of X_{an} in the site $X_{1h} = \mathrm{Top}/X_{an}$, and so there is a natural continuous morphism of sites $X_{et} \to X_{1h}$. It can be shown[230], essentially as a consequence of 12.7.1, that this morphism induces an isomorphism on cohomology groups with finite coefficients.

One can also show[231] that the Verdier homotopy type of the site X_{1h} is naturally isomorphic to the singular homotopy type[232] of X_{an}. Let H* denote ordinary cohomology for manifolds. Putting the above remarks together we have the following.

<u>Theorem 12.7.2</u> *If* X *is a smooth scheme of finite type over* C *and* X_{an} *is its associated complex manifold, then for any group* **A** *there is a natural homomorphism* $H*(X_{et};A) \to H*(X_{an};A)$ *which is an isomorphism if* **A** *is finite. In particular, for any prime* ℓ *there is a natural isomorphism* $H*_{et}(X;Q_\ell) \to H*(X_{an};Q_\ell)$.

Before we can state the other comparison theorems, we need a further definition.

<u>Definition 12.7.3</u>   A map $f:X \to Y$ of schemes is <u>proper</u> if it is finite type and for any map $g:Z \to Y$ the fibre product $X \times_Y Z \to Z$ is a closed map (between the base spaces).

Since the map $P^n(Z) \to Z$ is proper[233] and proper maps are preserved by base change, it follows that if V is a projective K-variety then the structure map $sch(V) \to K$ is proper.   The second comparison theorem is as follows[234].

<u>Theorem 12.7.4</u>   *Suppose* X *is a proper scheme over* k *and* K/k *is an extension with both* k,K *separably algebraically closed.   The natural map* $X_{et} \to (X \times_k K)_{et}$ *of sites induces an isomorphism on cohomology with finite coefficients.   In particular, if* ℓ *is a prime, then there is a natural isomorphism* $H^*_{et}(X;Q_\ell) \to H^*_{et}(X \times_k K;Q_\ell)$.

The third comparison theorem[235] links lifting and reduction of varieties.

<u>Theorem 12.7.5</u>   *Suppose* K *is the algebraic closure of* Q *, and* S *is a discrete spot on* K *with valuation ring* R *and residue field* k.   *If* X *is a proper scheme over* R *then the natural maps* $X_{et} \to (X \times_R K)_{et}$ *and* $X_{et} \to (X \times_R k)_{et}$ *induce isomorphisms on cohomology whose coefficient groups are finite of order prime to* $char(k)$.

12.8   COHOMOLOGICAL PROOFS OF THE WEIL CONJECTURES

Let $\underline{V_{sp}(K)}$ denote the category of smooth projective varieties over the algebraically closed field K, and let ℓ be a prime.   Provided ℓ is not the characteristic of K then the functor $H^*_{et}(-;Q_\ell)$ from $\underline{V_{sp}(K)}$ to the category of graded Q_ℓ-algebras has all the properties of a cohomology theory that we used in Chapter 10.   For example, $H_{et}^r(V;Q_\ell)$ is finite

dimensional[236] over Q_ℓ and is zero for $r > 2\dim(V)$ and $H_{et}^*(-,Q_\ell)$

satisfies Poincaré duality[237] and the Lefschetz fixed point formula.[238]

Suppose that V is a smooth projective variety defined over the finite

field F_q and $\pi:V \to V$ is the Frobenius map.   Put $P_r(t) = \det(1 - tH_{et}^r(\pi;Q_\ell))$

which is a polynomial over Z_ℓ.   It follows as in Chapter 10 that

$$Z_V(t) = \prod_{r=0}^{2\dim(V)} P_r(t)^{(-1)^{r+1}}$$

and so $Z_V(t)$ is a rational function of t over Q_ℓ, and satisfies the

functional equation 9.12.2 by virtue of Poincaré duality (compare with §10.5).

Suppose K is the algebraic closure of Q and S is a discrete spot on K

with valuation ring R and residue field k of characteristic p.   The field k

is algebraically closed by 4.3.8.   If X is a smooth proper scheme over R

such that $X \times_R k = \mathrm{sch}(V)$,   then if ℓ is a prime not equal to p and

$A = Z/\ell^n Z$ we have

$$
\begin{aligned}
H^*(V_{et};A) &= H^*((X \times_R k)_{et};A) \\
&= H^*((X \times_R K)_{et};A) & \text{by 12.7.5} \\
&= H^*((X \times_R C)_{et};A) & \text{by 12.7.4} \\
&= H^*(X_{an};A) & \text{by 12.7.2}
\end{aligned}
$$

and so $H_{et}^*(V;Q_\ell)$ is naturally isomorphic to $H^*(X_{an};Q_\ell) = H^*(X_{an};Q) \otimes Q_\ell$.

Thus the degree of the polynomial $P_r(t)$ (which is the dimension of $H_{et}^r(V;Q_\ell)$)

is precisely the r-th Betti number of the complex manifold X_{an}, which

confirms conjecture (9.12.5).   The independance of $P_r(t)$ on the prime ℓ

was proved by Deligne[239].

The proof of the Riemann hypothesis is however a much more formidable

task as we have seen even in the complex case (see Chapter 11).   The proof

of this conjecture by Deligne[240] uses cohomology with coefficients in sheaves.

198

Footnotes

1. See Cashwell and Everett (31), Gilmer (63).

2. See Sansone and Gerretsen, vol.1 (162), §7.2.1, p.353.

3. Ibid., §7.1.3, p.351.

4. Ibid., §7.1.2, p.351.

5. Two holomorphic functions defined in half-planes are _equivalent_ if they agree in some half-plane, and a _germ_ is an equivalence class.

6. This may be proved directly, as in Hardy & Wright (86), §17.1, p.245.

7. See Titchmarsh (194), §9.42, p.300.

8. See Titchmarsh (195), Theorem 2.1, p.13.

9. Ibid., pp.1,2, or Hardy & Wright (86), Theorem 280, p.246.

10. See Sansone and Gerretsen, vol.1 (162), §3.11.1, p.156.

11. In his investigations into the number of primes (see Riemann (155) p.148), Riemann says of the function
$$\xi(t) = \Gamma(s)(s-1)\zeta(s)\pi^{-t/2} \qquad \text{where } s=\tfrac{1}{2}+ti$$

    "Man findet nun in der That etwa so viel reele Wurzeln innerhalb dieser Grenzen und es ist sehr wahrscheinlich dass alle Wurzeln reell sind. Hiervon wäre allerdings eine strenge Beweis zu wanschen; ich habe indess die Aufsuchung desselben nach einigen flüchtigen vergeblichen Veruschen vorläufig bei Seite gelassen, da er für nächsten Zweck meiner Untersuchung entbehrlich sind."

12. See Hardy and Wright (86), Chapter 17 for further details and examples.

13. For the definition and properties of integral closure, see Atiyah and Macdonald (17), Chapter 5, pp.59-73.

14. See Cohn (36), Theorem 3, p.96 for example.

15. See Cohn (36), Chapter 6, §9, p.110, or Niven and Zuckermann (148), Chapter 7, in particular §7.8, pp.175-179.

16. See Atiyah and Macdonald (17), Chapter 9, pp.93-98.

17. This zetafunction was introduced by Dedekind in 1871 (see Dedekind (37)), who proved convergence in H_1. Landau proved that ζ extends to a meromorphic function in the half plane H_a where a = 1 - deg(K/Q)$^{-1}$ (see Landau (116)), and Hecke finally proved the extension to C, see Hecke (94). See Narkiewicz (146), Chapter VII, Prop.7.1, p.293 and Theorem 7.1, p.296.

18. See Narkeiwicz (146), Chapter VI, Prop.7.1, p.293. The class number relation was first proved by Dedekind (see Dedekind (37)).

19. This was first proved by Hecke in Hecke (94).

20. See Ribbenboim (154), pp.97-98.

21. See Artin (6).

22. It is easy to show from the equation $1 + X = (\Sigma \binom{\frac{1}{2}}{r} X^r)^2$ by equating coefficients inductively that $4^r \binom{\frac{1}{2}}{r} \in Z$.

23. See Artin (6), pp.196-197.

24. Ibid., p.180 (imaginary case) and p.194 (real case).

25. See Sansone and Gerretsen (162), §4.1, pp.175-176.

26. See Artin (6), p.207.

27. Ibid., p.209.

28. Ibid., p.216.

29. Ibid., p.223.

30. See §4.7 and Armitage (5).

31. This follows from a simple calculation using the formula.

32. See Artin (6), equation (5), p.209.

33. See example 8.3.2.

34. For the definition and properties of transcendence degree, see Winter (217), p.41.

35. See Weil (214), Chapter III, §2, lemma 1, p.48.

36. See Endler (58), §1.14, p.5 and §1.15, p.6.

37. Ibid., §2,p.8.

38. Ibid., §1.5, p.2.

39. Immediate from the definition.

40. See 4.5.5.

41. See Artin (9), Theorem 7, p.37.

42. Ibid., Chapter 1, §6, pp.43-44.

43. Ibid., Theorem 12, p.47.

44. Ibid., Chapter 2, Theorem 1, p.53 and Theorem 3, p.54.

45. A non-trivial discrete subgroup G of R^* is clearly generated by any non-trivial element $g \in G$ at which $|\log(g)|$ attains its minimum.

46. See Serre (172), Prop.8, p.44.

47. The ideals $A_n = m_S^n A$ for $n \in N$ have trivial intersection. For $a \in A$ non-zero put $|a| = 2^{-n}$ where $n = \max\{n : a \in A_n\}$. This norm defines a topology on A making it a topological R_S-algebra.

48. See Endler (58) (16.7), pp.120-122, or Zariski and Samuel (220), Vol.II, Theorem 17, p.279.

49. See Atiyah and Macdonald (17), Prop.2.6, p.21.

50. See Artin (9) for example.

51. Of course there are many choices of embedding $Q^a \subset C$.

52. In particular, Dwork's proof of rationality of the zeta function of a variety over a finite field (Dwork (55,56)) uses this technique. This will also be relevant in Chapter 12.

53. See Artin (9), Chapter 1, Theorem 9, p.45.

54. See Artin and Whaples (10).

55. See O'Meara (150), Theorem 11.8, p.8.

56. See note 22.

57. See Serre (178), Theorem 4, p.18.

58. This is a consequence of the Riemann-Roch theorem. See
     Eichler (57), pp.133-135.

59. See Weil (214), pp.4-8.

60. See Weil (214), pp.8-12, or Pontrjagin (152), §2.7,
     Theirem 22, pp.170-171.

61. See O'Meara (150), Theorem 21.2, p.42.

62. Ibid., §22B, pp.45-46.

63. Ibid., Theorem 22.1, p.46.

64. Ibid., Theorem 33.10, p.77.

65. Ibid.

66. See Atiyah and Macdonald (17), pp.96-98 for the definition
     of fractional ideals and their properties.

67. A module M over the ring A is locally-free of rank n if for
     each prime ideal P of A, the localisation M_P is a free A_P
     module of rank n. See Atiyah and Macdonald (17), Chapter
     3, pp.36-49.

68. This follows essentially from (5.2.1.2).

69. For the restricted product (also called local direct pro-
     duct), see Hewitt and Ross, Vol.I (96), §23.33, pp.373-374
     or Bracconier (25).

70. This is not true for every locally-compact abelian group.

71. See Weil (214), Chapter 4, pp.59-79, Narkiewicz (146),
     Chapter 6, Roberts (158) or Tate (192).

72. See Hewitt and Ross, Vol.I (96), §23.33, pp.373-374, §23.27,
     pp.366-367, §25.1, pp.400-402 and §25.2, pp.402-403.

73. See Weil (214), Theorem 6, p.76.

74. This proof is essentially that given in Weil (214) or
     Tate (192).

75. See Mordell (143).

76.   See Halmos (85), Theorem D, p.110.

77.   This is well known when K_S is the real or complex field,
       the character being a suitable power of the exponential
       map. Otherwise, take any character whose restriction to
       R_S is the composite of the reduction map $R_S \rightarrow R_S/M_S \rightarrow K(S)$
       with a non-trivial character on $K(S)$.

78.   See Weil (214), Corollary 1, p.69.

79.   Ibid., Prop.6, p.113.

80.   See Hewitt and Ross (96), Vol.II, §31.46e, p.246 and
       Mordell (143).

81.   The field K_S is itself a field of Laurent series by results
       of §4.5.

82.   Choose elements $X, Y \in E$ such that $E/k(X)$ is a finite separ-
       able extension generated by Y, then replace Y by a suitable
       multiple by some element of $k(X)$ so that Y is integral over
       $k[X]$.   The curve defined by the minimum polynomial of Y
       has E as its field of fractions.

83.   In fact $A[Y]/M$ is isomorphic to $K(S)$.

84.   See Atiyah and Macdonald (17), Prop.5.13, p.63.

85.   This follows from Shafarevich (181), Theorem 3, p.111.

86.   See §9.9 for a precise definition.

87.   See Atiyah and Macdonald (17), Theorem 7.5, p.81.

88.   First proved by Schmidt (164). The algebraic proofs given
       for example in Semple and Kneebone (165) or Walker (200)
       are equally valid in characteristic non-zero.

89.   See Shafarevich (181), Chapter 2, §4, pp.120-123.

90.   To be more precise, one starts with a model having only
       ordinary singularities, which one can do by virtue of
       Noether's Theorem. See Semple and Kneebone (165), Theorem
       11, p.133 and Notes 28,29, pp.280-281.

91.   For a constructive proof of Puisceux's Theorem, see Semple
       and Kneebone (165), pp.340-346 or Walker (200), Theorem 3.1,
       pp.98-102.

92.   See Chevalley (33), p.64.

93.   See Artin (9), Chapter 2, §7, pp. 47-52.

94.   See Milnor (140), pp. 13-14.

95.   See Cartan (28), Theorem 3, p.180.

96.   The construction of meromorphic functions on a compact
      Riemann surface is usually via automorphic functions on its
      universal covering space. Such functions can be construc-
      ted as the quotient of two automorphic forms of the same
      weight. See Kra (113), pp.119-124.

97.   See Sansone & Gerretsen (162), Vol.1, p.66.

98.   This is true for Riemann surfaces since it is true for C.

99.   An immediate corollary of Liouville's theorem.

100. An 'index theorem' relates invariants of different struc-
      tures on a manifold - in these cases the algebraic and
      analytic structures. See Hirzebruch (99), pp.1,2 and
      184-196 or Palais (151).

101. Strictly speaking we need only know that S^2 can be triangu-
      lated since we can always lift a suitably fine triangula-
      tion of S^2 to one of X using the covering space property.
      This is one way of proving that Riemann surfaces can be
      triangulated.

102. See Spanier (183), Theorem 14, p.172.

103. See Macdonald (134), §14 p.331 for example, or Martens
      (137) Chapter 2, §2,3 and Mattuck (139).

104. Clearly if X $= S^n$ for n $\geq$ 3 then $X^{(2)}$ is not a manifold.

105. See §8.1, p.107 for the definition.

106. See Chow (35).

107. See Spanier (183), Corollary 3, p.80, Theorem 2, p.86.

108. Ibid., Theorem 7, p.195. For if S^2 is a topological group,
      then for any x $\in S^2$ the map a $\to$ ax is homotopic to the
      identity and so has a fixed point which is impossible.

109. Ibid., result 12, p.149.

110. It is a locally-compact group, and is finite if X is com-
      pact of genus at least 2.

111.   If $f : \tilde{X} \to X$ is the universal covering space of X,
        $G \subseteq \mathrm{Aut}(\tilde{X})$ is the group of covering transformations and N
        is the normalizer of G then the natural map $N \to \mathrm{Aut}(\tilde{X}/G) =$
        Aut(X) induces an isomorphism $N/G \to \mathrm{Aut}(X)$. By the Rie-
        mann classification theorem (see note 112), $\tilde{X} = H$, C or S^2.
        Explicit calculation (see note 113) shows that $\mathrm{Aut}(\tilde{X})$ is a
        Lie group, hence so is N (being a closed subgroup), and
        hence so is Aut(X).

112.   See Springer (184) Chapter 9, especially Theorem 9.1
        and the following remarks, pp.224-225.

113.   See Cartan (28), Chapter 6, §2.2, 2.3 and 2.4, pp.178-181.

114.   This is a simple exercise in canonical forms.

115.   The lattice L is determined uniquely up to homothety, i.e.
        multiplication by a non-zero scalar.

116.   See Lang (120), pp.10-11.

117.   Ibid., Chapter 3, §3, pp.39-41.

118.   See Hasse (90) III, p.202.

119.   Ibid., §2, pp.200-203.

120.   See §9.11, p 139.

121.   Ibid.,§2-3, p.203.

122.   See Note 87.

123.   See Mumford (144), Theorem 1, p.11.

124.   See Atiyah & Macdonald (17), Chapter 4, pp.50-58.

125.   See Weil (203), p.1.

126.   Ibid., Chapter II, pp.26-44.

127.   See Shafarevich (181), Theorem 1, p.110.

128.   Ibid., Theorem 3, p.111.

129.   See Godement (65), Chapitre II, §1.2, pp.110-112.

130.   Ibid., Theorème 1.2.1, p.111.

131.   Gaga also refers specifically to Serre (167).

132.   For manifolds, the analogous condition is often called
        <u>transverse</u> intersection.

133.   See Serre (179), Chapter V, §C.

134.   See Kleiman (108), Griffiths (69), Grothendieck (80),
        Roberts (159), Samuel (161).

135.   See Roberts (159).

136.   If $f(X) = X^q - X$ then the equation $f(X) = 0$ has no multiple
        roots since $df/dx = 1$. Thus the fixed points of π in
        affine space, and hence in V (since V is locally affine)
        have multiplicity one.

137.   When V is projective, f(Y) is already closed, since pro-
        jective varieties are complete, and complete varieties play
        the role of compact spaces. See Shafarevich (181), Theorem
        2, p.45.

138.   Ibid., pp.154-155.

139.   See Chow (35).

140.   See Weil (207) and Weil (211).

141.   Rationality was first proved by Dwork (see 55,56) or Serre
        (175), using analytic methods over the complex and p-adic
        numbers simultaneously, in particular using an SMR (see
        §4.6). For the other conjectures, see §12.7.

142.   See Whitney (216), pp.301-304 and Artin (13).

143.   See Chow (34) or Griffiths & Adams (70), Chapter 3, §3,
        pp.69-72.

144.   See Greenberg (68), Spanier (183), Chapter 6, pp.292-306,
        or Vick (199), Chapter 5, pp.124-172.

145.   See Vick (199), Theorem 3.9, p.86.

146.   See Lojasewicz (131).

147.   See Heller & Rowe (95) or Godement (65) pp.127-136.

148.   See Tennison (193), Theorem 3.4, p.132.

149.   See Godement (65), p.133, Tennison (193), Theorem 6.9, p.52.

150.   See Godement (65), Chapter 3, §4,5 in particular Theorem
        4.7.1, p.181, and Example 7.2.1 p.263, and Hirzebruch (99),

§2.12, pp.35-36.

151. See Godement (65), Chapter 2, §3.7, pp.156-158.

152. Ibid., §3.1, pp.147-149 (flabby = flasque).

153. See Spanier (183), Exercises D1-3, pp.358-359.

154. For the definition of a pro-object in a category, and the corresponding Pro-category, see Artin & Mazur (16), Appendix pp.147-166.

155. See Spanier (183), Theorem 10, pp.335-336.

156. Once we have chosen an orientation for C^n this induces an orientation on each n-dimensional complex manifold via the charts, which is unique since holomorphic maps of C^n are orientation preserving.

157. See Vick (199), Proposition 5.30, p.161 or Spanier (183), Theorem 18, p.297.

158. See Borel & Haefliger (23), §4.15, p.488, or Atiyah & Hirzebruch (18) for the analogous results with singular cohomology.

159. See Borel & Haefliger (23), §4.11, §4.12, p.486.

160. When V is a curve, every map f:V → V other than the identity has only finitely many fixed points.

161. One reason for working with cohomology theories for sheaves is that there is a natural concept of theories with a family of supports. See Godement (65) or Swan (188).

162. For example, it is an isomorphism when X is a Grassmanian. See Grothendieck (72).

163. See Lefschetz (127) and Brown (26).

164. See Whitney (216), pp.176-177 or Hodge & Pedoe (103), pp.286-292.

165. An exercise in cyclotomic polynomials.

166. This is because $G_{r,s}(C)_{an}$ has the structure of a CW-complex with cells only in even dimensions.

167. The Poincaré polynomial of $GL_n(C)$ is $\prod_1^n (1 + X^{2i+1})$ and so it

follows from a theorem of Leray that the Poincaré polynomial of the homogeneous space $G_{r,s}$ is $g_{r+s}(X)/g_r(X)g_s(X)$, where $g_r(X) = \prod_1^r (1-X^{2i})$, which is precisely $f(X^2)$.

168. See for example Carlitz (27), Hua & Vandiver (104), Lang & Weil (123) and Macdonald (134).

169. There is an algebraic proof of this. See Neron (147).

170. See Mitchell (141), Theorem 4.1, p.251, Tennison (193), Proposition 7.6, p.55 and Proposition 7.10, p.56.

171. See Swan (187).

172. See Weil (203), Theorem 1, p.13.

173. See Whitney (216), pp.157-162 and Kobayashi & Nomizu (109), pp.30-32.

174. See Grothendieck & Dieudonne (84), Chapter 0, Theorem 20.4.8 or Mumford (144), p.279.

175. See Flanders (60), §3.2, pp.20-22.

176. See Fossum (61), p.27.

177. See Hirzebruch (99), §3.8, p.49.

178. See Grothendieck (73).

179. See Hirzebruch (99), Theorem 2.12.2, p.37.

180. See Hirzebruch (99), Theorem 2.11.2, p.35.

181. Ibid., Theorem 2.12.3, p.37 and Flanders (60), Chapter 5, especially §5.9, pp.66-69.

182. See Kobayashi & Nomizu (109), Volume 1, Chapter 4, especially pp.154-155.

183. See Flanders (60), §5.8, pp.64-66.

184. See Flanders (60),§8.4, pp.136-139 and Hodge (101).

185. First proved by Grothendieck. See Dolbeault (53).

186. See Hirzebruch (99), p.46.

187. See Kra (113), Chapter 2, §1, pp.43-46.

188. The proof of this is analogous to that of the existence of
     a Riemannian metric. See note 182.

189. The proof of this is similar to that of the real case:
     see note 184.

190. See Hirzebruch (99),pp.121, 184-186, Kodaira (110) and
     Palais (151).

191. See Kobayashi & Nomizu (109), Volume 1, Example 1.2, p.155.

192. See Hirzebruch (99), p'123, Kähler (106), Weil (212).

193. The group Z acts freely as a group of holomorphic isomor-
     phisms of the complex manifold $C^2\setminus\{0\}$ by $n.(z_1,z_2) =$
     $(2^n z_1, 2^n z_2)$. The quotient space with the induced complex
     structure is topologically $S^1 \times S^3$, which has $\beta_1 = 1$, so
     does not admit a Kähler metric.

194. See Serre (171).

195. See Weil (204), Théorème 8, p.42 and Théorème 10, p.54.

196. See Eichler (57), Chapter 4, pp.281-299.

197. See Weil (204), pp.20-21.

198. Ibid., Corollaire 2, p.69.

199. See Spanier (183), Chapter 6, exercise E1-6, pp 359-360.

200. See Martens (137) or Griffiths (69).

201. See Martens (137).

202. See Weil (211).

203. See Lang (120), pp.172-178 or Swinnerton-Dyer (190),pp.68-
     86.

204. The nullstellensatz for spec R is trivial in comparison
     with the nullstellensatz for affine varieties (see 9.1.2.),
     and is simply a restatement of the fact that the radical of
     an ideal is the intersection of all prime ideals containing
     it. See Atiyah & Macdonald (17), Proposition 1.14, p.9.

205. See Atiyah & Macdonald (17), Exercise 26, pp.14-15.

206. See Macdonald (135), Proposition 5.1, p.37.

207. Ibid., p.52.

208. See Tennison (193), Exercise 5, p.109.

209. See Mumford (144), pp.156-158.

210. Ibid., pp.165-180.

211. This is often taken as the definition of a variety.

212. See Grothendieck-Dieudonne (84), no.4, Proposition 6.1.4, p.141 and no.8, Proposition 1.5.4, p.234.

213. See Mumford (144), pp.181-199.

214. See Grothendieck-Dieudonne (84), no.32, 17.1.1, 17.1.3 and 17.3.1, pp.56-61.

215. See Mumford (144), Corollary 1, p.356.

216. See Artin, Grothendieck & Verdier (14), Tome 1, Exp.II, III, pp.219-298.

217. Ibid., Exp.III, §5, pp.293-297.

218. Ibid., pp.311-312.

219. Ibid., Exp.II, §5, pp.251-257 and Exp.IV, pp.350-353.

220. See Artin (11), pp.30-34.

221. Ibid., Miscellany 1.8 (iii), p.33.

222. Ibid., Theorem 1, pp.24-30.

223. See Artin, Grothendieck & Verdier (14), Tome II, Exp.V, §7, pp.49-80 and Artin-Mazur (16).

224. See Artin, Grothendieck & Verdier (14), Tome 1, Exp.II, Proposition 4.8, p.249.

225. See Artin, Grothendieck & Verdier (14), Tome II, Théorème 7.4.1, pp.62-63.

226. See Grothendieck & Dieudonné (84), no.4, Corollary 6.1.9, p.142.

227. See Deuring (50), §8, pp.251-258.

228. See Lang (120), pp.172-174 or Serre (177).

229.   See Serre (176), Theorem 1, Grauert-Remmert (66), Theorem
        1 and (67), Theorem 4, and Grothendieck (76), pp.10-11.

230.   See Artin & Mazur (16), Theorem 12.9, pp.142-143, Artin,
        Grothendieck & Verdier (14), Tome III, Exp.XVI, §4, pp.
        232-245.

231.   This follows from Artin & Mazur (16), Corollary 12.2, p.
        129 and Artin, Grothendieck & Verdier (14), Tome 1, Exp.IV
        pp.358-362.

232.   The singular homotopy type of a space X is the simplicial
        set S^*X where S^nX is the set of singular n simplices in X
        (see Spanier (183), p.160), regarded as an object of HSS,
        which is a subcategory of Pro-HSS.

233.   See Mumford (144), Theorem 2, p.238.

234.   See Artin, Grothendieck & Verdier (14), Tome III, Exp.XII,
        Corollary 5.4, pp.89-90.

235.   See Artin & Mazur (16), Corollary 12.13, p.144.

236.   See Artin (11) in the case of curves, Grothendieck (77)
          §3 and Artin, Grothendieck & Verdier (14) in general.

237.   See Expose XVIII by Deligne in Artin, Grothendieck Verdier
        (14), Tome III, pp. 481-587, especially Théorème 3.2.5,
        p. 585.

238.   See Grothendieck (77).

239.   See Deligne (44) Théorème 1.6 and Verdier (198).

240.   See Deligne (42,44) and Serre (180).

Bibliography

1.   Abhyankar, S.S., Tame coverings and fundamental groups of
      algebraic varieties, Amer.J.Math. 81 (1959) 46-94.

2.   Abhyankar, S.S., Local analytic geometry, Academic Press 1964.

3.   Abhyankar, S.S., On the problem of resolution of singulari-
      ties, Proc.Int.Cong.Math. (1966) 469-481.

4.   Altman, A. & Kleiman, S., Introduction to Grothendieck
      duality, Lecture Notes in Mathematics 146,Springer
      1970.

5.   Armitage, J.V., On the genus of curves over finite fields,
      Mathematika 9 (1962) 115-117.

6.   Artin, E., Quadratische Körper im Gebiet der höhren
      kongruenzen, Math.Zeit., 19 (1924) 153-246.

7.   Artin, E., Geometric algebra, Interscience, 1957.

8.   Artin, E., Collected papers, Addison-Wesley, 1965.

9.   Artin, E., Algebraic numbers and algebraic functions,
      Gordon and Breach, 1968.

10.   Artin, E. & Whaples, G., Axiomatic characterisation of
      fields by the product formula for valuations,
      Bull.Amer.Math.Soc. 51 (1945) 469-492.

11.   Artin, M., Grothendieck topologies, mimeographed notes,
      Harvard 1962.

12.   Artin, M., The étale cohomology of schemes, Proc.Int.Cong.
      Math. (1966) 44-56.

13.   Artin, M., "The implicit function theorem in algebraic
      geometry" in Algebraic geometry (S.S.Abhyankar et
      al) 13-34, Oxford Univ.Press, 1969.

14.   Artin, M., Grothendieck, A. & Verdier, J.L., Théorie des
      topos et cohomologie étale des schémas (SGA 4),
      3 vols., Lecture Notes in Mathematics 269, 270,
      305, Springer 1972-73.

15. Artin, M. & Mazur, B., "Homotopy of varieties in the étale
         topolgy", in Proc.Conf. on local fields (T.A.
         Springer, ed.), 1-15, Springer 1967.

16. Artin, M. & Mazur, B., Etale homotopy, Lecture Notes in
         Mathematics 100, Springer 1969.

17. Atiyah, M.F. & Macdonald. I.G., Introduction to commutative
         algebra, Addison-Wesley, 1969.

18. Atiyah, M.F. & Hirzebruch, F., Analytic cycles on complex
         manifolds, Topology 1 (1962) 25-45.

19. Barr, M., Grillet, P.A. & Osdol. D.H. van, Exact categories
         and categories of sheaves, Lecture Notes in
         Mathematics 236, Springer 1971.

20. Berthelot, P., Cohomologie cristalline des schémas en
         caractéristique p > 0, Lecture Notes in Mathematics
         407, Springer 1974.

21. Berthelot, P., Grothendieck, A. & Illusie, L., Théorie des
         intersections et théorème de Riemann-Roch (SGA 6)
         Lecture Notes in Mathematics 225, Springer 1971.

22. Bombieri, E., Counting points on curves over finite fields,
         Sem.Bourbaki 430, Lecture Notes in Mathematics
         383, Springer 1974.

23. Borel, A., & Haefliger, A., La classe d'homologie fondamen-
         tale d'une espace analytique, Bull.Soc.Math.
         France 89 (1961) 461-513.

24. Borevich, Z.I. & Shaferevich, I.R., Number theory, Academic
         Press, 1966.

25. Braconnier, J., Sur les groupes topologiques localement
         compacts, J.Math.Pures Appl. 27 (1948) 1-85.

26. Brown, R.F., The Lefschetz fixed-point theorem, Scott,
         Foresman and Co., 1971.

27. Carlitz, L., Pairs of quadratic equations in a finite field,
         Amer.J.Math. 76 (1954) 137-154.

28. Cartan, H., Elementary theory of analytic functions of one
         or several complex variables, Hermann, 1963.

29. Cartan, H., Variétés Riemanniennes, variétés analytiques
         complexes, variétés Kähleriennes, Sem.H. Cartan
         1951-52 exp.1, W.A. Benjamin, 1967.

30. Cartan, H. & Chevalley, C., Géometrie algébrique, Séminaire
         de l'Ecole Normal Supérieure 8e année (1955-56).

31. Cashwell, E.D. & Everett, C.J., The ring of number theoretic
         functions, Pacific J.Math. 9 (1959) 975-985.

32. Cassels, J.W.S. & Fröhlich, A. (ed.), Algebraic number
         theory, Academic Press, 1967.

33. Chevalley, C., Algebraic functions of one variable, Math.
         Surveys 6, Amer.Math.Soc., 1951.

34. Chow, W.L., On compact complex analytic varieties, Amer.J.
         Math. 71 (1949) 893-914.

35. Chow, W.L., The Jacobian variety of an algebraic curve,
         Amer.J.Math. 76 (1954) 453-476.

36. Cohn, H., A second course on number theory, J. Wiley 1962.

37. Dedekind, R., "Über die theorie der ganzen algebräischen
         Zahlen", 297-314, Gesammelte mathematische
         Werke, Vol.III, Chelsea 1969.

38. Dedekind, R. & Weber, H., Theorie der algebräischen Funktio-
         nen einer Veränderlichen, J. Reine Angew.Math.
         92 (1882) 181-290.

39. Deligne, P., Théorie de Hodge I, Proc.Int.Cong.Math. 1970
         Vol.I, 425-430.

40. Deligne, P., Théorie de Hodge II, Publ.Math. IHES 40 (1971)
         5-58.

41. Deligne, P., Théorie de Hodge III, Publ.Math. IHES 44 (1974).

42. Deligne, P., La conjecture de Weil pour les surfaces K3,
         Invent.Math. 15 (1972) 206-226.

43. Deligne, P., Poids dans la cohomologie de variétés algébriques,
         Proc.Int.Cong.Math (1974), Vol.I, 79-85.

44. Deligne, P., La conjecture de Weil, I, Publ.Math. IHES 43
         (1974), 273-307.

45. Deligne, P. & Katz, N., Groupes de monodromie en géometrie
         algébrique (SGA 7II), Lecture Notes in Mathematics
         340, Springer 1973.

46. Demazure, M., Motifs des variétés algébriques, Sem.Bourbaki
         365, Lecture Notes in Mathematics 180,Springer 1971.

47.   Demazure, M. & Grothendieck, A., Schémas en groupes (SGA 3)
      (3 vols.), Lecture Notes in Mathematics 151, 152
      153, Springer 1970.

48.   de Rham, G., Variétés différentiables, Hermann 1955.

49.   Deuring, M., Arithmetische theorie der Korrespondenzen
      algebräischer funktionenkörper, J.Reine Angew.
      Math. 177 (1937) 161-191 and 183 (1941) 25-36.

50.   Deuring, M., Die typen der multiplikatorenringe elliptischer
      funktionenkörper, Abh.Math.Sem.Univ. Hamburg 14
      (1941) 197-272.

51.   Deuring, M., Die zetafunktion einer algebräischen kurve von
      geschlechte,Nachr.Akad.Wiss.Göttingen Math-Phys.,
      I (1953) 85-94, II (1955) 13-42, III (1956) 37-76,
      IV (1957) 55-80.

52.   Deuring, M. Lectures on the theory of algebraic functions
      of one variable, Lecture Notes in Mathematics
      314, Springer 1973.

53.   Dolbeault, P., Sur la cohomologie des variétés analytiques
      complexes, C.R.Acad.Sci. Paris 236 (1953) 175-177.

54.   du Val. P., Elliptic Functions and elliptic curves, Cambridge
      University Press 1973.

55.   Dwork, B., On the rationality of the zeta function of an
      algebraic variety, Amer.J.Math. 82 (1960) 631-648.

56.   Dwork, B., "Analytic theory of the zeta function of
      algebraic varieties", in Arithmetical Algebraic
      Geometry, O.F.G. Schilling (ed.) 18-32, Harper
      & Row, 1963.

57.   Eichler, M., Introduction to the theory of algebraic
      numbers and functions, Academic Press 1966.

58.   Endler, O., Valuation theory, Springer 1972.

59.   Fatou, P., Séries trigonométriques et séries de Taylor,
      Acta Math. 30 (1906) 335-400.

60.   Flanders, H., Differential forms, Academic Press 1963.

61.   Fossum, R.M., The divisor class group of a Krull domain,
      Springer 1973.

62.   Fulton, W., Algebraic curves, W.A. Benjamin, 1969.

63.   Gilmer, R., Power series over Krull domains, Pacific J.Math.
         29 (1969) 543-549.

64.   Giraud, J., "Analysis situs", in Dix exposés sur la coho-
         mologie des schémas (J. Giraud et al), North-
         Holland 1968.

65.   Godement, R., Théorie des faisceaux, Act.Sci. et Ind. 1252,
         Hermann 1964.

66.   Grauert, H. & Remmert, R., Espaces analytiquement complets,
         C.R. Acad.Sci. Paris 245 (1957) 882-885.

67.   Grauert, H. & Remmert, R., Komplex Räume, Math.Annalen
         136 (1958) 245-328.

68.   Greenberg, M.J., Lectures in algebraic topology, W.A.
         Benjamin 1973.

69.   Griffiths, P.A., "Some results on algebraic cycles on
         algebraic manifolds", in Algebraic Geometry,(S.S.
         Abhyankar et al) 93-199, Oxford Univ. Press 1969.

70.   Griffiths, P. & Adams, J., Topics in algebraic and analytic
         geometry, Princeton University Press 1974.

71.   Grothendieck, A., Sur quelques points d'algèbre homologique,
         Tohoku Math.J. 9 (1957) 119-121.

72.   Grothendieck, A., Sur quelques propriétés fondamentales en
         théorie des intersections, Sem. C. Chevalley,
         E.N.S., 1958.

73.   Grothendieck, A., La théorie des classes de Chern, Bull.Soc.
         Math. France 86 (1958), 137-154.

74.   Grothendieck, A., The cohomology theory of abstract algebraic
         varieties, Proc.Int.Cong.Math. (1958) 103-118.

75.   Grothendieck, A., On the de Rham cohomology of algebraic
         varieties, Publ.Math. IHES 29 (1966) 95-103.

76.   Grothendieck, A., Technique de descente et théorèmes
         d'existence en géometrie algébrique, Sem.Bourbaki
         190, W.A. Benjamin 1966.

77.   Grothendieck, A., Formule de Lefschetz et rationalité des
         fonctions L, Sem. Bourbaki 279, W.A. Benjamin 1966.

78.   Grothendieck, A., "Crystals and the de Rham cohomology of
         schemes", in Dix exposés sur la cohomologie des

schémas (J. Giraud et al), 306-358, North-
Holland, 1968.

79.   Grothendieck, A., Cohomologie locale des faisceaux cohérents
          et théoremes de Lefschetz locaux et globaux
          (SGA 2), North-Holland 1968.

80.   Grothendieck, A., "Standard conjectures on algebraic cycles"
          in Algebraic Geometry (S.S. Abhyankar et al),
          193-199, Oxford Univ. Press 1969.

81.   Grothendieck, A., Revêtements étales et groupe fondamental
          (SGA 1), Lecture Notes in Mathematics 224,
          Springer 1971.

82.   Grothendieck, A., Groupes de monodromie en géometrie
          algébrique (SGA 7I), Lecture Notes in Mathematics
          288, Springer 1972.

83.   Grothendieck, A., Cohomologie l-adique et fonctions L.
          (SGA 5) (mimeographed notes), IHES.

84.   Grothendieck, A. & Dieudonné, J., Elements de géometrie
          algébrique I,II,III,IV, Publ.Math. IHES, nos.
          4,8, (11,17), (20,24,28,32), 1960-1967.

85.   Halmos, P.R., Measure theory, Van Nostrand 1950.

86.   Hardy, G.M. & Wright, E.M., An introduction to the theory
          of numbers, Oxford Univ. Press, 1965.

87.   Hartshorne, R., On the de Rham cohomology of algebraic
          varieties, Publ.Math. IHES 45 (1975) 5-99.

88.   Hasse, H., Beweis analogous der Riemannschen Vermtung für
          die Artinsche und F.K. Schmidtschen Kongruenz-
          zetafunktionen in gewissen elliptischen Fallen,
          Nachr.Akad.Wiss.Göttingen (1933) 253-262.

89.   Hasse, H., Abstrakte begrundung der Komplexen multipli-
          takion und Riemannsche Vermutung in Funktionen-
          körpern, Abh.Math.Sem.Univ. Hamburg 10 (1934)
          325-348.

90.   Hasse, H., Zur theorie der abstrakten elliptische funktion-
          enkörper, J. Reine Angew.Math. 175 (1936), I
          55-62, II 69-88, III 193-208.

91.   Hasse, H., Zahlentheorie, Akademie-Verlag 1950.

92. Hasse. E., Mathematische Abhandlungen (3 vols.), Walter de Gruyter 1975.

93. Hecke, E., Über die Zetafunktion beliebiger algebräischer Zahlkörper, Nachr.Akad.Ges.Wiss.Göttingen (1917) 77-89.

94. Hecke, E., Mathematische Werke, Vandenhoeck & Ruprecht 1959.

95. Heller, A. & Rowe, K.A., On the category of sheaves, Amer. J.Math. 84 (1962) 205-216.

96. Hewitt, E. & Ross, K.A., Abstract Harmonic Analysis (2 vols) Springer 1963.

97. Hironaka, H., "On the equivalence of singularities I", in Arithmetical Algebraic Geometry (O.F.G. Schilling ed.), Harper & Row 1963, 153-200.

98. Hironaka, H., Resolution of singularities of an algebraic variety over a field of characteristic zero, I,II, Ann. of Math. 79 (1964), 109-326.

99. Hirzebruch, F., Topological methods in algebraic geometry, 3rd edition, Springer 1966.

100. Hodge, W.V.D., The geometric genus of a surface as a topological invariant, J. London Math.Soc. 8 (1933), 312-318.

101. Hodge, W.V.D., The theory and application of harmonic integrals, Cambridge, Univ. Press 1941.

102. Hodge, W.V.D., Some recent developments in the theory of algebraic varieties, J. London Math.Soc. 25 (1950), 143-157.

103. Hodge, W.V.D. & Pedoe, D., Methods of algebraic geometry, Cambridge Univ. Press 1952.

104. Hua, L. & Vandiver, H., Characters over certain types of rings with applications to the theory of equations in a finite field, Proc.Nat.Acad.Sci. U.S.A. 35 (1949), 94-99.

105. Iwasawa, K., Lectures on p-adic L functions, Ann.Math. Studies 74, Princeton Univ. Press 1972.

106. Kähler, E., Geometria Arithmetica, Ann.di Mat. 45 (1958) 1-368.

107. Katz, N. & Messing, W., Some consequences of the Weil
      conjectures, Invent.Math. 23 (1974), 73-77.

108. Kleiman, S., "Algebraic cycles and the Weil conjectures",
      in Dix exposés sur la cohomologie des schémas
      (J. Giraud et al), 359-386, North-Holland 1968.

109. Kobayashi, S. & Nomizu, K., Foundations of differential
      geometry (2 vols.), Interscience 1963.

110. Kodaira, K., Some results in the transcendental theory of
      algebraic varieties, Proc.Int.Cong.Math.(1954),
      Vol.III, 474-480.

111. Kodaira, K. & Spencer, D.C., Groups of complex line bundles
      over compact Kähler varieties, Proc.Nat.Acad.
      Sci.U.S.A. 39 (1953), 868-872.

112. Kodaira, K. & Spencer, D.C., Divisor class groups on
      algebraic varieties, Proc.Nat.Acad.Sci.U.S.A.
      39 (1953), 872-877.

113. Kra, I., Automorphic forms and Kleinian groups, W.A.
      Benjamin 1972.

114. Krull, W., Dimensiontheorie in Stellenring, J.Reine Angew.
      Math. 179 (1938), 204-226.

115. Lamotke, K., Semisimpliziale algebraische topologie,
      Springer 1968.

116. Landau, E., Ueber die zu einem algebraischen Zahlkörper
      gehorige Zetafunktion und die Ausdehnung de
      Tschebyscheftschen Prinzahlentheorie auf das
      Problem der Vertheilung des Primideals, J.Reine
      Angew.Math.125 (1903), 64-188.

117. Lang, S., Introduction to algebraic geometry, Interscience
      1958.

118. Lang, S., Algebraic number theory, Addison-Wesley 1970.

119. Lang, S., Introduction to algebraic and abelian functions,
      Addison-Wesley 1972.

120. Lang, S., Elliptic functions, Addison-Wesley 1973.

121. Lang, S., $SL_2(R)$, Addison-Wesley 1975.

122. Lang, S. & Serre, J.-P., Sur les revêtements non ramifiés des
      variétés algébriques Amer.J.Math.79 (1957) 319-330.

123. Lang, S. & Weil, A., Number of points of varieties over
      finite field s, Amer.J.Math.76 (1954), 819-827.

124. Lascoux, A. & Berger, M., Varietes Kähleriennes compactes,
      Lecture Notes in Mathematics 154, Springer 1970.

125. Lefschetz, S., On certain numerical invariants of algebraic
      varieties with applications to abelian varieties,
      Trans.Amer.Math.Soc. 22 (1921), 327-482.

126. Lefschetz, S., L'analysis situs et la géometrie algébrique
      Gauthier-Villars, 1924.

127. Lefschetz, S., Intersections and transformations of
      complexes and manifolds, Trans.Amer.Math.Soc.
      28 (1927), 1-49.

128. Lefschetz, S., Algebraic geometry, Oxford Univ.Press 1953.

129. Lichtenbaum, S., "Values of zeta-functions, etale coho-
      mology and algebraic K-theory", in Algebraic
      K-theory II, 489-501, Lecture Notes in
      Mathematics 342, Springer 1973.

130. Lieberman, D., Numerical and homological equivalence of
      algebraic cycles on Hodge manifolds, Amer.J.
      Math. 90 (1968), 366-374.

131. Lojasewicz, S., Triangulating semi-analytic sets, Anni.
      della scuolo Norm.Sup. di Pisa, Ser.III 18
      (1964), 449-474.

132. Lubkin, S., On a conjecture of André Weil, Amer.J.Math.
      89 (1967), 443-547.

133. Lubkin, S., A p-adic proof of Weil's conjectures, Ann. of
      Math. 87 (1968), 105-194, 194-255.

134. Macdonald, I.G., Symmetric products of an algebraic curve,
      Topology 1 (1962), 319-343.

135. Macdonald, I.G., Algebraic geometry: Introduction to
      Schemes, W.A. Benjamin Inc. 1968.

136. Manin, Y.I., Cubic forms:algebra, geometry and arithmetic,
      North-Holland 1974.

137. Martens, H.H., Three lectures on the classical theory of
      Jacobian varieties in Algebraic geometry Oslo
      1970 (F.Oort ed.) 141-170 Wolters-Noordhoff 1972.

138.   Matsushima, Y., Differentiable manifolds, Marcel Dekber
            1972.

139.   Mattuck, A., Symmetric products and Jacobians, Amer.J.
            Math. 83 (1961), 189-206.

140.   Milnor, J., Singular points of complex hypersurfaces,
            Annals of Maths. Studies no.61, Princeton Univ.
            Press 1968.

141.   Mitchell, B., Theory of categories, Academic Press 1965.

142.   Monsky, P. & Washnitzer, G., Formal cohomology I, Ann. of
            Math. 88 (1968) 181-217.

143.   Mardell, L.J., Poisson's summation formula and Riemann's
            zeta function, J. London Math.Soc. 4 (1929)
            285-291.

144.   Mumford, D., Introduction to algebraic geometry (mimeo-
            graphed notes), Harvard University.

145.   Nagata, M., Local rings, Interscience 1962.

146.   Narkeiwicz, W., Elementary and analytic theory of algebraic
            numbers, Monografie Matematyczne 1974.

147.   Neron, A., Problemes arithmétiques et géometriques
            rattachés à la notion de rang d'une courbe
            algébrique dans un corps, Bull.Soc.Math. France
            80 (1952) 101-166.

148.   Niven, I. & Zuckerman, H.S., An introduction to the theory
            of numbers (second edition), J. Wiley 1966.

149.   Noether, M., Zur Grundlegung der Theorie der algebräischen
            Raumkurven, J.Reine Angew.Math. 93 (1882) 271-318.

150.   O'Meara, O.T., Introduction to quadratic forms, Springer
            1973.

151.   Palais, R.S., Seminar on the Atiyah-Singer index theorem,
            Princeton Univ. Press 1965.

152.   Pontrjagin, L., Topological groups (second ed.), Gordon
            and Breach 1966.

153.   Popp, H., Fundamentalgruppen algebräischer Mannigfaltig-
            keiten, Lecture Notes in Mathematics 176,
            Springer 1970.

154.   Ribbenboim, P., Algebraic numbers, Wiley-Interscience 1972.

155.   Riemann, B., Sur les nombres premiers inférieurs à une
          grandeur donné, Monatsberichte der Berliner
          Akadamie (1859), 671-680.

156.   Riemann, B., Gesammelte Mathematische Werke (ed. H.Weber),
          Dover 1953.

157.   Robert, A., Elliptic curves, Lecture notes in Mathematics
          326, Springer 1973.

158.   Robert, A., Des adèles:pourquoi? L'Enseignement math. 20
          (1974), 133-145.

159.   Roberts, J., "Chow's moving lemma", in Algebraic geometry
          Oslo 1972 (F. Oort ed.), 89-96, Wolters-
          Noordhoff 1972.

160.   Sampson, S.H. & Washnitzer, G., Numerical equivalence and
          the zeta-function of a variety, Amer.J.Math. 81
          (1959), 735-748.

161.   Samuel, P., Relations d'equivalence en géometrie algébrique
          Proc.Int.Cong.Math. (1958), 470-487.

162.   Sansone, G. & Gerretsen, J., Lectures on the theory of
          functions of a complex variable (2 vols.),
          P. Noordhoff 1960.

163.   Schilling, O.F.G., The theory of valuations, Math.
          Surveys 4, Amer.Math.Soc. 1950.

164.   Schmidt, F.K., Analytische Zahlentheorie in Körperen der
          charakteristic p, Math.Zeit. 33 (1931) 1-32.

165.   Semple, J.G. & Kneebone, G.T., Algebraic curves, Oxford
          Univ. Press 1959.

166.   Serre, J.-P., Faisceaux algébriques cohérents (FAC),
          Ann. of Math. 61 (1955) 197-278.

167.   Serre, J.-P., Géometrie algébrique et géometrie analytique
          (GAGA), Ann. Inst. Fourier 6 (1956) 1-42.

168.   Serre, J.-P., Sur la cohomologie des variétés algébriques,
          Journ. de Math. 36 (1957) 1-16.

169.   Serre, J.-P., "Sur la topologie des variétés algébriques
          en caractéristique p", in International Sympo-
          sium on Algebraic Geometry, Universidad Nacional

de Mexico 1958, 24-53.

170.   Serre, J.-P., Groupes algébriques et corps de classes,
          Act.Sci. et Ind. 1264, Hermann 1959.

171.   Serre, J.-P., Analogues Kähleriens de certains conjectures
          de Weil, Ann. of Math. 71 (1960) 392-394.

172.   Serre, J.-P., Corps locaux, Act.Sci. et Ind. 1296,
          Hermann 1962.

173.   Serre, J.-P., "Zeta and L functions", in Arithmetical
          Algebraic Geometry (O.F.G. Schilling, ed.),82-92
          Harper & Row 1963.

174.   Serre, J.-P., Théorie du corps de classes pour les revête-
          ments non ramifiés de variétés algébriques,
          Sem. Bourbaki 133, W.A. Benjamin 1966.

175.   Serre, J.-P., Rationalité des fonctions ζ des variétés
          algébriques, Sem. Bourbaki 198, W.A.Benjamin 1966.

176.   Serre, J.-P., Revêtements ramifiés du plan projectif, Sem.
          Bourbaki 204, W.A. Benjamin 1966.

177.   Serre, J.-P., Abelian l-adic representations and elliptic
          curves, W.A. Benjamin 1968.

178.   Serre, J.-P., A course in arithmetic, Springer 1973.

179.   Serre, J.-P., Algèbre locale. Multiplicités (Third edition),
          Lecture Notes in Mathematics 11, Springer 1975.

180.   Serre, J.-P., Valeurs propres des endomorphismes de
          Frobenius, Sem. Bourbaki 446, Lecture Notes in
          Mathematics 431, Springer 1975.

181.   Shaferevich, I.R., Basic algebraic geometry, Springer 1974.

182.   Shimura, G. & Taniyama, Y., Complex multiplication of
          abelian varieties and applications to number
          theory, Publ.Math.Soc. Japan, 1961.

183.   Spanier, E.H., Algebraic topology, McGraw-Hill 1966.

184.   Springer, G., Introduction to Riemann surfaces, Addison-
          Wesley 1957.

185. Stepanov, S.A., On the number of points of a hyperelliptic
      curve over a finite prime field (Russian), Izv.
      Akad.Nauk.SSSR.Ser.Mat. 33 (1969) 1171-1181.

186. Sullivan, D., Geometric topology (mimeographed notes),
      Massachusetts Institute of Technology, 1970.

187. Swan, R.G., Vector bundles and projective modules, Trans.
      Amer.Math.Soc. 105 (1962), 264-277.

188. Swan, R.G., The theory of sheaves, University of Chicago
      Press 1964.

189. Swinnerton-Dyer, H.P.F., "An outline of Hodge theory", in
      Algebraic geometry Oslo 1970 (S.S. Abhyankar et
      al), 277-286, Wolters-Noordhoff, 1972.

190. Swinnerton-Dyer, H.P.F., Analytic theory of abelian
      varieties, Cambridge Univ. Press 1974.

191. Tate, J.T., "Algebraic cycles and poles of zeta functions"
      in Arithmetical Algebraic Geometry (O.F.G.
      Schilling ed.), 95-109, Harper & Row 1963.

192. Tate, J., "Fourier analysis in number fields and Hecke
      zeta functions", in Algebraic number theory,
      (J.Cassels & A.Frohlich eds.), pp.
      Academic Press 1967.

193. Tennison, B.R., Sheaf Theory, Cambridge Univ.Press 1975.

194. Titchmarsh, E.C., The theory of functions, Oxford Univ.
      Press 1932.

195. Titchmarsh, E.C., The theory of the Riemann zeta function,
      Oxford Univ. Press 1951.

196. Tuskina, T.A., A numerical experiment on the calculation
      of the Hasse invariant for certain curves, Amer.
      Math.Soc.Transl.(2) 66 (1968), 204-205.

197. Verdier, J.-L., "The Lefshetz fixed point formula in étale
      cohomology", in Proc. Conf. on local fields
      (T.A. Springer ed.) 199-214, Springer 1967.

198. Verdier, J.-L., Indépendance par rapport à ℓ des polynomes
      caractéristiques des endomorphismes de Frobenius
      de la cohomologie l-adique, Sem.Bourbaki 423,
      Lecture Notes in Mathematics 383, Springer 1974.

199.   Vick, J.W., Homology theory:an introduction to algebraic
        topology, Academic Press 1973.

200.   Walker, R.J., Algebraic curves, Princeton Univ.Press 1950.

201.   Weil, A., Sur les fonctions algébriques à corps de con-
        stants finis, C.R.Acad.Sci. Paris 210 (1940),
        592-594.

202.   Weil, A., On the Riemann hypothesis in function fields,
        Proc.Nat.Acad.Sci. U.S.A. 27 (1941) 345-347.

203.   Weil, A., Foundations of algebraic geometry, American Math-
        metical Society Colloq.Publ. XXIX 1946.

204.   Weil, A., Sur les courbes algébriques et les variétés qui
        s'en deduisent, Act.Sci.et Ind. 1041, Hermann
        1948.

205.   Weil, A., Variétés abeliennes et courbes algébriques,
        Act.Sci.et Ind. 1064, Hermann 1948.

206.   Weil, A., On some exponential sums, Proc.Nat.Acad.Sci.
        U.S.A. 34 (1948) 204-207.

207.   Weil, A., Number of solutions of equations in finite fields
        Bull.Amer.Math.Soc. 55 (1949) 497-508.

208.   Weil, A., Number theory and algebraic geometry, Proc.Int.
        Cong.Math. (1950), Vol.II 90-100.

209.   Weil, A., Jacobi sums as grossencharakter, Trans.Amer.
        Math.Soc. 73 (1952) 487-495.

210.   Weil, A., Footnote to a recent paper, Amer.J.Math. 76
        (1954) 347-350.

211.   Weil, A., Abstract versus classical algebraic geometry,
        Proc.Int.Cong.Math.(1954),Vol.III 550-558.

212.   Weil, A., Introduction à l'étude des variétés Kähleriennes
        Hermann 1958.

213.   Weil, A., "Zeta functions and Mellin transforms", in
        Algebraic geometry (S.S. Abhyankar et al) 409-
        426, Oxford Univ. Press 1969.

214.   Weil, A., Basic number theory, Springer 1972.

215.   Weir, A.J., Notes on algebraic geometry, Queen Mary College
        Mathematics Notes.

216. Whitney, H., Complex analytic varieties, Addison-Wesley 1972.

217. Winter, D.J., The structure of fields, Springer 1974.

218. Zariski, O., The fundamental ideas of abstract algebraic geometry, Proc.Int.Cong. Math.(1950) Vol.II, 77-89.

219. Zariski, O., Theory and applications of holomorphic functions on algebraic varieties over arbitrary ground fields, Mem. Amer. Math.Soc. 5 (1951).

220. Zariski, O.& Samuel, P., Commutative algebra, 2 volumes, Van Nostrand 1958-60.

221. Zimmer, H.G., Some computational aspects of, and the use of computers in, algebraic number theory, Computing 8 (1971) 363-381.

Additional biography to the second edition.

Griffiths,P. & Adams. J., Topics in algebraic and analytic geometry, Princeton University Press 1976.

Hartshorne, R. Algebraic geometry, Springer 1977.

Katz, N.M., "An overview of Deligne's proof of the Riemann Hypothesis for finite fields", in Proc.Symp. in Pure Mathematics Vol.28, 275-305,Amer.Math.Soc. 1976.

Mumford, D.,Algebraic geometry I,complex projective varieties,Springer 1976.